I0390695

DIFFERENTIAL EQUATIONS WITH MATHEMATICAL MODELLING

AND ANALOG PROGRAMMING

BY

NKEMJIKA CHINEDU B.

PREFACE

Differential equations with its use in
mathematical modeling of systems and
analog Programming is made simple in this
book. The book is intended to teach
differential equations and their use in
mathematical modeling and analog
programming in a simple way. I have
written this book for students and
teachers whose works are on this area and
also for readers who may have interest in
this area of study. Examples, solved
problems and exercises to test their
skills, and understanding, I have
included for their use.

Nkemjika Chinedu B.

Table Of Content

Chapters:

Chapter One

A differential equation is an equation that comprises of one or more derivatives or differentials. It consists of derivatives of various quantities. *We can also describe it as an equation relating an unknown function with one or more of its derivatives.* This form of equation arises from a situation in which change is occurring. Typical examples of differential equations are:

$d^2x/dt^2 + 4dx/dt + 2x = 3t$ ... (1.)

Or $dq/dt + 5q = sint$ ... (2.)

where for equation (1.) X, the variable being differentiated, the dependent variable and t is the independent variable. Similarly q is the dependent variable for equation (2) and t the independent variable. Solutions to these differential equations may be analytical or where we have difficult problems we use numerical methods to obtain approximate solutions.

Differential equations may be classified by the following:

(a) **Type**:

here we have ordinary and partial differential equations. It is ordinary if the dependent variable, say y, depends on only a single independent variable, say x, that is y = f(x), y is a function of only one single independent variable, x. It is however partial if the dependent variable, a single dependent variable, say y, depends on two or more independent variables say x and t that is y = f(x , t), that is the value of y is determined by the value of x and t.

(b) **Order:**

the order of a differential equation is the order or number of its highest derivative. Examples are:

(i) d^2y/dt^2 + dy/dt + 6y = 0

the highest derivative in the above equation is two and thus the order is two. It is hence called a second order differential equation.

(ii) dy/dt + 6y = 5

the highest order derivative in the above equation is one and thus it is a first order differential equation. The order of a differential equation, note, will determine the number of arbitrary constants in the solution of the equation. A differential equation of n order has n arbitrary constants.

(c) **Degree**: this is the exponent of the highest power of the highest order derivatives after the equation has been cleared of fractions and radicals in the dependent variable and its derivatives. Example:
$(d^3y/dt^3)^2 + (d^2y/dt^2)^5 +$
$(y/(x^2 + 1)) = e^x$
the derivative, from above, with the highest order is d^3y/dt^3 and its power (exponent) is 2. Thus the degree of the equation is two.

The solution of a differential equation:

the solution to a differential equation is a differentiable function that satisfies the differential equation within an interval (a , b) of values for the independent variables. A differential equation often has different functions that satisfy it, or many solutions. The general solution, a solution involving arbitrary constants, comprises all these different functions and from this general solution one can get all possible solutions through varying the arbitrary constants. Application of given conditions to the general solution yields a particular solution. The number of the given independent conditions, note, must be the same as the number of arbitrary constants for one to obtain a particular solution.Examples for illustration are given below:

example 1:

verify that $y = 3\sin2x$ is a solution of $(d^2y/dx^2) + 4y = 0$.

Solution:

$(d^2y/dx^2) = -4y$

Since $y = 3\sin2x$ then $(d^2y/dx^2) =$ $-4(3\sin2x) = -12\sin2x$

Now differentiating the function $y = 3\sin2x$ twice to get a second derivative of y will give:

$dy/dx = -6\cos2x$

$d^2y/dx^2 = -12\sin2x = -4(3\sin2x) = -4y$

Hence $y = 3\sin2x$ is a solution.

Example 2:

verify that $y = A\cos x + B\sin x$ satisfies the differential equation $(d^2y/dx^2) + y = 0$.

Verify also that $y = A\cos x$ and $y =$

Bsinx each individually satisfies the equation.

$(d^2y/dx^2) = -y$

Since $y = A\cos x + B\sin x$

then $(d^2y/dx^2) = -(A\cos x + B\sin x)$,

now $y = A\cos x + B\sin x$

$dy/dx = -A\sin x + B\cos x$,

$(d^2y/dx^2) = -A\cos x - B\sin x$

$= -(A\cos x + B\sin x)$

hence $y = A\cos x + B\sin x$ satisfies the differential equation.

For $y = A\cos x$:

$(d^2y/dx^2) = -y = -A\cos x$.

Now since $y = A\cos x$

$dy/dx = -A\sin x$,

$(d^2y/dx^2) = -A\cos x$

hence $y = A\cos x$ satisfies the equation $(d^2y/dx^2) + y = 0$.

For $y = B\sin x$ where $(d^2y/dx^2) + y = 0$:

$(d^2y/dx^2) = -y$

Hence with $y = B\sin x$

$(d^2y/dx^2) = -B\sin x$.

Now since $y = B\sin x$

$dy/dx = B\cos x$

$(d^2y/dx^2) = -B\sin x$

Hence $y = B\sin x$ satisfies the differential equation when $y = B\sin x$.

Example 3:

if $y = Ae^{2x}$ is the general solution of $dy/dx = 2y$, find the particular solution satisfying y(0) = 3. What is the particular solution satisfying dy/dx = 2 when x = 0 ?

at y(0) = 3

$y = Ae^{2x}$ and

y(0) = 3

$= Ae^{2(0)}$

$= Ae^0 = A$

thus A = 3.

Hence $y = Ae^{2x} = y = 3e^{2x}$.

The particular solution satisfying dy/dx = 2 when x = 0 will be: getting the first derivative of $y = Ae^{2x}$ that is $dy/dx = 2Ae^{2x}$ hence for dy/dx = 2 at x = 0 we have for

$dy/dx = 2Ae^{2x}$

$= 2 = 2Ae^{2(0)}$

$= 2Ae^0 = 2A$

Thus $2 = 2A$

$A = 1$

We thus have the particular solution as $y = e^{2x}$.

Example 4:

the general solution of the equation $(d^2x/dt^2) - 3(dx/dt) + 2x = 0$ is given by $x = Ae^t + Be^{2t}$ find the particular solution which satisfies $x = 3$ and $dx/dt = 5$ when $t = 0$.

for x = 3 at t = 0 we have:

(d^2x/dt^2) - 3dx/dt + 2x = 0 where general solution is x = Ae^t + Be^{2t}

we have 3 = $Ae^{(0)}$ + $Be^{2(0)}$,

3 = A + B ... (1)

for dx/dt = 5 at t = 0 we have

from x = Ae^t + Be^{2t}

dx/dt = Ae^t + $2Be^{2t}$

hence at dx/dt = 5 and t = 0 we have : 5 = $Ae^{(0)}$ + $2Be^{2(0)}$,

5 = A + 2B ... (2).

Solving for A and B simultaneously we have:

3 = A + B

5 = A + 2B

Equation(2) - equation(1): 2 = B and substituting B = 2 in either equation(1) or equation(2) we have using equation(1):

3 = A + 2

A = 3 - 2 = 1

Thus the particular solution:

x = e^t + 2e^{2t}

Exercises:

1.   The   general   solution   of
(d^2y/dx^2) - 2dy/dx + y = 0 is

$Y = Axe^x + Be^x$. Find the particular solution satisfying $y(0) = 0$, $(dy/dx)(0) = 1$.

(Solution: $y = xe^x$.)

2. Verify that $3e^x$, Axe^x, $Axe^x + Be^x$, where A, B are constants, all satisfy the differential equation: $d^2y/dx^2 - 2(dy/dx) + y = 0$.

3. Verify that $x = t^2 + A\ln t + B$ is a solution of $(td^2x/dt^2) + dx/dt = 4t$

Given a differential equation our interest is to find all of its solutions. Hence our intention will be focused on finding mainly the general solution to different types of differential equations.

1. ORDINARY DIFFERENTIAL EQUATIONS:

a. First order:

(i) Simple differential equations: this is the simplest first order differential equation. The form is

$$dy/dx = f(x)$$

where the function f on the right hand side depends only on the independent variable (x). Solution to this simple equation is obtained by direct integration. Examples are given below:

(i.) $dy/dx = 3$

Solution:

multiplying both sides by dx :

(dy/dx)*dx = 3dx ,

$\int$ (dy) =

$\int$ (3dx), y = | ((3x$^{0 + 1}$)/(0 + 1))+

C =   3x + C

(ii.) dy/dt = 6t

multiplying both sides by dt :

(dy/dt)*dt = 6t*dt

$\int$ (dy) = $\int$ (6tdt)

y = |((6t$^{(1 + 1)}$)/(1 + 1))+ c =

(6t^2)/2 + C =   3t^2 + C.

1. Find the general solution of the following equations:

(a) dx/dt = 5

(b) dy/dx = 2x

(c) dy/dx = $8x^2$

(d) dx/dt = $3t^3$

(2) find the particular solution of the following equations :

(a) dx/dt = 3t , x(0) = 1.

(b) dx/dt = Int , x(1) = 1.

(c) dx/dt = t^2, x(0) = 2.

(1a.) x = 5t + C.

(b) y = x^2 + C.

(c) y = (8/3)x^3 + C

(d) x = (3/4)t^4 + C.

(2a.) x = (3/2)t^2 + 1.

(b) tIn|t|-t + C, tIn|t| - t + 2.

(c) x = ((t^3)/3)+ 2.

(ii) *Separable differential Equations:*

these are first order equations that can be written in the form:
(dy/dx) = f(x)g(y),
the function on the right hand side depends on both the dependent and the independent variables.

Solution to separable equations can be obtained by collecting like terms, y term with its dy term and x term with its dx term, separately together on the different sides of the equation and then integrating them accordingly. Examples are provided below:

(a) $dy/dx = (x^2)/y$

collect like terms together:

$ydy = x^2dx$

then integrating:

$$\int (ydy) = \int (x^2dx)$$

| $(y^{(1 + 1)})/(1 + 1) =$

| $((x^{2 + 1})/(2+1)) + C$

$(y^2)/2 = ((x^3)/3) + C$

$y^2 = 2(((x^3)/3) + C)$

$Y = \sqrt{(2(((x^3)/3) + C))}$

(b) $dy/dx = e^{-2x}/y^2$

Solution:

---

collecting like terms we have:

$y^2dy = e^{-2x}dx$

Integrating both sides:

$\int (y^2dy) = \int (e^{-2x}dx)$

$| (y^{2+1}/(2+1)) = |((e^{-2x})/-2) + C$

$y^3/3 =$

$|((e^{-2x})/-2)+ C = ((e^{-2x})/-2)+ C$

(c) $dy/dx = e^{x-y}$

Solution:

$dy/dx = e^x \div e^y = e^x/e^y$

$dy/dx = e^x/e^y$

Collecting like terms:

$e^y dy = e^x dx$

$\int (e^y dy) = \int (e^x dx)$

$| e^y = | e^x + C$

$e^y = e^x + C$

separate the variables and solve
the following differential
equations:

(1) $(\sqrt{(2xy)}).(dy/dx) = 1$

(2) $\ln x (dx/dy) = (x/y)$

(3) $x(2y - 3)dx + (x^2 + 1)dy = 0$

(4) $(\sqrt{(1 + x^2)})dy +$
$(\sqrt{(y^2 - 1)})dx = 0$

(5) $9(dy/dx) = 2x$

Solution:

(1) $y^{(3/2)} = 3(\sqrt{(x/2)}) + c.$

(2) $(1/2)(\ln|x|)^2 = \ln|y| + C.$

(3) $(x^2 + 1)(2y - 3) = C.$

(4) $\cosh^{-1}y + \sinh^{-1}x = C.$

(5) $y = 1/9(x^2 + C)$

(iii) *Exact differential equations:*

an equation that can be solved simply by integrating both sides of the equation is called an exact equation. Example:

(a) $dy/dx = 2x^2$

, $(dy/dx) * dx = (2x^2)dx$

$$\int (dy) = \int (2x^2dx).$$

$y = ((2x^{2 + 1})/(2 + 1)) + C$

$y = ((2x^3)/3) + C$

(b) $d(xy)/dx = 2x^2$

multiplying through by dx we have:

$(d(xy)/dx) * dx = 2x^2 dx$

$$\int (d(xy)) = \int (2x^2 dx)$$

$xy = | ((2x^{2+1})/(2+1)) + C$
$= ((2x^3)/3) + C.$

dividing through by x :

$xy/x = (1/x)((2x^3/3) + C) =$
$y = ((2x^3)/3x) + (C/x) =$
$(2x^2x/3x) + Cx^{-1} = ((2x^2)/3) + (C/x)$

One can easily identify or recognise an exact equation. This is because an exact equation takes the form below:

$u(dy/dx) + u'y = f(x)$

Where

(1): u is a function of x.

(2.): the coefficient of y (u') is a derivative of the coefficient of dy/dx (that is u).

Hence under the above conditions the left hand side of the equation, that is , $u(dy/dx) + u'y$ can be written as $d(uy)/dx$ (*product rule in differentiation*). This will thus give $d(uy)/dx = f(x)$.

Examples:

each of the equations below is exact. Solve them.

(i) $x^2(dy/dx) + 2xy = x^3$

$(d(x^2y)/dx) = x^3$

$d(x^2y) = x^3dx$

$\int (d(x^2y)) = \int (x^3dx)$

$x^2y = (x^{3+1}/3 + 1) + C$

$x^2y = x^4/4 + C$

dividing through by x^2 :

$(x^2y/(x^2)) = ((x^4)/4) + C)(1/x^2)$

$= y = (((x^4)/4x^2) + C/x^2) =$

$(x^2/4) + Cx^{-2}$

(ii) $1/x^2(dy/dx) - (2/x^3)y = 5x^3$

Verify it is exact :

$d(1/x^2)/dx = d(x^{-2})/dx$

$= -2x^{-2-1} = -2x^{-3}$

hence $d((1/x^2)y)/dx = 5x^3$ is true.

Multiplying $d((1/x^2)y)/dx = 5x^3$ through by dx:

$(d((1/x^2)y)/dx)dx = 5x^3 dx$

$d((1/x^2)y) = 5x^3dx$

$\int (d(y/x^2)) = \int (5x^3dx)$

$y/x^2 = ((5x^{3+1})/ 3+1) + C$

$= ((5x^4)/4) + C ,$

$(y/x^2) * (x^2) =$

$[((5x^4)/4) + C] * (x^2) =$

$y = [(((5x^4)/4)x^2) + Cx^2]$

$= 5x^6/4 + Cx^2 .$

1. Each of the following equations is exact. Solve them. (a) $x^2(dy/dx) + 2xy = x^3$.

(b) $((1/x^2)dy/dx) - (2/x^3)y = 5x^3$

(c) $x^4(dy/dx) + 4x^3y = 4x$

(a) $y = ((x^2)/4) + C/(x^2)$.

(b) $y = ((5x^6)/4) + Cx^2$

(c) $(2/x^2) + C/x^4$

IV. *LINEAR DIFFERENTIAL EQUATIONS:*

a differential equation is considered linear if the dependent variable fulfill the following conditions:

i. the dependent variable and its derivative exist only in the first power.

ii. there are no products involving the dependent variable with its derivatives.

iii. there are no non-linear functions of the dependent variable such as sine and exponential of the dependent variable.

The first order linear equation can be written in the standard form:

dy/dx + P(x)y = Q(x)
where P and Q can only be functions of the independent variable x . The P or Q can also be a simple constant.
The first order linear equations, even when they are not exact equation, can be made exact by multiplying them through by an integrating factor. Consider the example below using the standard form equation:

dy/dx + P(x)y = Q(x) ,
here we want to make this equation exact by multiplying both sides of it by an integrating factor. Let this integrating factor be *u*.
Hence multiplying both sides of the equation by *u gives,*

$u(dy/dx) + uP(x)y = uQ(x)$,

this equation is now exact and as such the left hand side can be written as d(uy)/dx that is

$u(dy/dx) + uP(x) = d(uy)/dx$

since it can now be solved simply by integrating directly both sides of the equation. This implies that since

$u(dy/dx) + uP(x)y = uQ(x)$ then

$d(uy)/dx = uQ(x)$ and hence we can simply integrate both sides to solve, that is ,

$(d(uy)/dx)dx = uQ(x)dx$

$d(uy) = uQ(x)dx$,

$$\int (d(uy)) = \int (uQ(x)dx) ,$$

$$uy = \int (uQ(x)dx) ,$$

thus $(uy)/u = (1/u)*(\int (uQ(x)dx))$

$=$

$$y = 1/u(\int (uQ(x)dx))$$

u *is* the integrating factor and it can be obtained as below:

given that $d(uy)/dx =$
$u(dy/dx) + uPy$
we differentiate using product rule the L.H.S (left hand side) to get
$u(dy/dx) + y(du/dx)$
this hence implies that
$u(dy/dx) + y(du/dx) = d(uy)/dx$
$= \quad u(dy/dx) + uPy$
that is
$u(dy/dx) + y(du/dx) =$
$u(dy/dx) + uPy$
hence
$y(du/dx) = u(dy/dx) - u(dy/dx) +$
uPy ,
$y(du/dx) = uPy$,
$du/dx = uP$.
Note P is a function of x and hence solving by method of

separation of variables we have:

dx(du/dx) = uPdx ,

(1/u)du = (uPdx)/u ,

(1/u)du = Pdx

integrating both sides we have:

$$\int ((du/u)) = \int (Pdx)$$

thus we have

$$\text{In}u = \int (Pdx) ,$$

$$u = e^{\int Pdx}$$

the constant of integration has
been purposely omitted in this
case.

Examples:

(a) find the general solution of the following equations:

(1) $dx/dt + 6x = 4$

(2) $dy/dx - 6y = 9$.

(B) find the general solution of $dx/dt = 2x + 4t$. What is the particular solution which satisfies $x(1) = 2$?

Solution:

1. $dx/dt + 6x = 4$,

$P(x) = 6$ and $Q(x) = 4$

$u = e^{\int P(t)dt}$

where $P(t) = 6$

hence

$u = e^{\int 6dt}$ =

$e|6t^{0+1}/(0+1) = {}_e6t$

multiplying thro by the integrating factor we have:

$e^{6t}dx/dt + 6xe^{6t} = 4e^{6t}$

hence we have:

$$e^{6t}x = \int (4e^{6t}dt) =$$

$| \ (4e^{6t})/6$

$e^{6t}x = 2/3(e^{6t}) + C$

dividing through by e^{6t} to get x we have:

$e^{6t}x/(e^{6t}) =$

$2/3(e^{6t}*(1/e^{6t})) + C(1/e^{6t})$

$x = 2/3 + Ce^{-6t}$.

2. $(dy/dx) - 6y = 9$

Solution:

where $P(x) = -6$, $Q(x) = 9$

we find the integrating factor:

$u = e^{\int P(x)dx}$

since x is the independent variable here.

$u = e^{\int (-6dx)}$

$= e^{|-6x^{0+1}}/0 + 1 =$

$e^{|-6x} = e^{-6x}$.

Next we multiply through by an integrating factor:

$e^{-6x}dy/dx - 6y(e^{-6x}) = 9e^{-6x}$

we thus have:

$e^{-6x}y = \int (9e^{-6x}dx)$,

$e^{-6x}y = 9(\int (e^{-6x}dx)) =$

$9|e^{-6x}/-6$

$= 9e^{-6x}/-6 + C$,

$e^{-6x}y = -3/2(e^{-6x}) + C$

dividing through by e^{-6x} we have:

$e^{-6x}y(1/e^{-6x}) = -3/2((e^{-6x})(1/e^{-6x}))+$ $C(1/e^{-6x})$

$= -3/2 + Ce^{6x}$,

$y = -3/2 + Ce^{6x}$.

(b) *Solution:*

dx/dt = 2x + 4t ,

(dx/dt) - 2x = 4t   ,

P(t) = -2 ;

Q = 4t

thus

$$u = e^{\int P(t)\,dt}$$

$$= e^{\int(-2dt)} =$$

$$u = e^{\int(-2dt)}$$

$$_e|(-2t^{0+1})/0 + 1$$

$$= e^{|-2t} = e^{-2t} .$$

Multiplying through by the
integrating factor we have:

$(e^{-2t}(dx/dt)) - 2xe^{-2t} = 4te^{-2t}$

Since equation is now exact we
then have:

$$e^{-2t}x = \int (4te^{-2t}dt) ,$$

$$e^{-2t}x = 4(\int (te^{-2t}dt))$$

using integration by part method:

let u = t ; dv/dt = e^{-2t} ;

(du/dt)= 1,

(dv/dt)dt = e^{-2t}dt

dv = e^{-2t}dt

$\int$ (dv) = $\int$ (e^{-2t}dt)

v   = |e^{-2t}/-2 = e^{-2t}/-2

hence $\int$ (u(dv/dx)dx) =

uv - $\int$ (v(du/dx)dx) =

(e^{-2t}/-2)t - $\int$ (((e^{-2t})/-2)1*dt) =

(e^{-2t}/-2)t + 1/2($\int$ (e^{-2t})dt)

= t(e^{-2t}/-2) + 1/2|((e^{-2t})/-2),

e^{-2t}x =

4{t((e^{-2t})/-2) + 1/2((e^{-2t})/-2)}

= e^{-2t}x =

$(4/-2)t(e^{-2t}) + (4/2)\{(e^{-2t}/-2)\}+ C,$

$e^{-2t}x = -2te^{-2t} - e^{-2t} + C$

dividing through by e^{-2t} :

$e^{-2t}x/e^{-2t} =$

$(-2te^{-2t}/e^{-2t}) - (e^{-2t}/e^{-2t}) +$

(C/e^{-2t}) ,

$x = -2t - 1 + Ce^{2t}$

where $x(1) = 2$ we have

for $x = 2$ at $t = 1$:

$x(1) = -2(1) - 1 + Ce^{2(1)} =$

$2 = -2 - 1 + Ce^{2(1)} = 2,$

$-3 + Ce^{2(1)} = 2$,

$Ce^{2(1)} = 2 + 3$,

$Ce^{2(1)} = 5$,

$5 = Ce^2$,

thus $C = 5(1/e^2)$,

$C = 5e^{-2}$

this implies that

$x = -2t - 1 + ((5e^{-2})*(e^{2t}))$

$= -2t - 1 + 5e^{(-2+2t)}$

$$= -2t - 1 + 5e^{(2t-2)}$$
$$= -2t - 1 + 5e^{2(t-1)}.$$

Exercises.

---

(1) Find the general solution of the following equations:

(a) $(dy/dx) + y = 1$.

(b) $dy/dx + 2y = 6$

(c) $(dy/dx) - 3y = 2$.

(d) $dx/dt = 3x - 8$

(e) $(dy/dx) + 6y = x^2$.

(2) solve $x(dy/dx) + y = x^4$

(3) Find the particular solution of the following equations:

(a) $(dx/dt) - x = 4$, $x(0) = 2$.

(b) $dy/dx = 4y - 8$, $y(1) = 2$.

44

Solution:

1(a). $y = 1 + Ce^{-x}$.

(b) $y = 3 + Ce^{-2x}$.

(c) $y = Ce^{3x} - (2/3)$.

(d) $x = (8/3) + Ce^{3t}$.

(e) $y = ((x^2)/6) - (x/8) + (1/108) + (C/(e^{6x}))$.

(2.) $(x^4)/5 + (C/x)$

(3a.) $x = 6e^t - 4$.

(3b) $y = 2$.

(V) Homogenous equations:

equations that can be put in the
form,

$(dy/dx) = F(y/x)$

is said to be homogenous. These
equations which have variables
that cannot be separated can be
solved by transforming them into
equations whose variables can be
separated. This transformation can
be achieved by the introduction of
a new independent variable

$v = y/x$

hence since $v = y/x$,

$y = vx.$

Differentiating y with respect to
x will give

$dy/dx = d(vx)/dx$

using the product rule we have:

$v(dx/dx) + x(dv/dx) = v + x(dv/dx)$

thus since $(dy/dx) =$

$d(vx)/dx = v + x(dv/dx)$

then

dy/dx = F(y/x) = F(v)

can be written as

dy/dx = F(v) = v + x(dv/dx)

the above equation can now be

solved by the separation of

variables to get,

- F(v) + v + x(dv/dx)= 0

dividing through by 1/dv we have

(-F(v) + v)/dv = -x/dx

thus

(dv/(- F(v) + v)) = - dx/x

this is same as

0 = (dx/x) + dv/(-F(v) + v)

we then obtain after solving the

above the solution of the original

equation by replacing v with y/x.

Example:

Solve the homogenous equation below:

$(x^2 + y^2)dx + xydy = 0$

Solution:

make dy/dx subject of the equation, that is:

$(x^2 + y^2)(dx/dx) = -xy(dy/dx)$,

$(x^2 + y^2)/1 = -xy(dy/dx)$

divide through by $-xy$:

$((x^2 + y^2)/1) * (1/(-xy)) = dy/dx$

thus $dy/dx = (x^2 + y^2)/(-xy)$

divide numerator and denominator of RHS (that is $(x^2 + y^2/-xy)$ by x^2 to get the variables in either (y/x) or x/y form:

$dy/dx = ((x^2/x^2) + (y^2/x^2))/(-xy/x^2)$

$= (1 + ((y/x)^2))/(-y/x)$

let v = y/x then we have:

dy/dx = $(1 + v^2)/-v$ = F(v)

hence implying using the equation

0 = (dx/x) + (dv/(v - F(v)))

we have

=

(dx/x)+ (dv/(v - $((1 + v^2)/-v)$))

=   (dx/x) + (dv/((v/1)

+ $((1 + v^2)/v)$))

= $dx/x + dv/((v^2 + 1 + v^2)/v)$ =

(dx/x) + dv/($(2v^2 + 1)$/v) =

(dx/x) + ((dv/1) * (v/($2v^2$ + 1))) =

(dx/x)+ (vdv/($2v^2$ + 1))

solving the above equation by

method of integration we have:

let u = $2v^2$ + 1 then

du/dv = 4v

du = 4vdv

(du/4) = vdv

hence equation

(dx/x)+ (vdv/($2v^2$ + 1))

can be written as,

$= (dx/x) + ((du/4)/u) =$

$dx/x + du/4u = dx/x + 1/4(du/u)$

hence we have by taking (dx/x) to
the other side and by integration:

$-(dx/x) = 1/4(du/u)$

$\int (-dx/x) = \int (1/4(du/u))$

$= -Inx + Inc = 1/4(Inu)$,

$Inx + (1/4)Inu = Inc$

since integration of

$\int (dx/x) = Inx$

applying the rule of logarithm we
have $(nlogP = logP^n)$

$Inx + Inu^{(1/4)} = Inc$,

multiplying all by 4 and applying
the rule of logarithm we have

$Inx^4 + Inu = Inc^4$

now since $logP + logQ = logPQ$

then we have

In$(x^4 u)$ = Inc^4.

Taking exponential on both sides and cancelling out we have:

$_e$In$(x^4 u)$ = $_e$Inc^4 = C

Since e^{Inx} = x we then have

$x^4 u$ = C =

$x^4 (2v^2 + 1)$ = C

since u = $2v^2 + 1$

hence since v = (y/x) we have

$x^4((2(y^2/x^2))+ 1)$ = C

$X^4(2(y^2/x^2) + 1)$ = C

$x^4((2y^2 + x^2)/x^2)$ = C

$x^2(2y^2 + x^2)$ = C.

Exercise:

(1) Solve the homogenous equations
:

(a) xdy - 2ydx = 0.   (B) (x + y)dy + (x - y)dx = 0.

Solution:

(a) $x^2 = yC$. (b) $2\tan^{-1}(y/x) +$
$\text{In}(x^2 + y^2) = C$

2. SECOND ORDER LINEAR EQUATIONS

the general form of a second order
linear ordinary differential
equation is
$p(x)(d^2y/dx^2) + m(x)(dy/dx) + k(x)y$
$= f(x)$ .... (1.)
where $p(x)$, $m(x)$, $k(x)$ and $f(x)$
are functions of x only.
If $f(x) = 0$ that is if the
function that is independent of y
is ignored we have the equation in
the form
$p(x)(d^2y/dx^2) + m(x)(dy/dx) + k(x)y$
$= 0$ ... (2.)
this equation is called a
homogenous equation, all its term
contains y or its derivative. The
other equation however, that is
equation (1.) With $f(x)$, is called
nonhomogenous.

The properties below are necessary for finding the solution of a second order linear differential equations -

Property1:

i. If $y_1(x)$ and $y_2(x)$ are two linearly independent solutions of a second order homogenous equation, that is one solution is not a simple multiple of the other, the general solution, $y_H(x)$ is:
$y_H(x) = Ay_1(x) + By_2(x)$
where A and B are arbitrary constants. Hence the second order ordinary linear differential equation has two arbitrary constants in its solution.

Property2:

let $y_p(x)$ be any solution of the
nonhomogenous equation and $y_H(x)$
the general solution of a
corresponding or associated
homogenous equation then the
general solution of a
nonhomogenous equation is

$y(x) = y_H(x) + y_p(x)$

where $y_H(x)$ is called the general
solution of an associated
homogenous equation or a
complimentary solution.

$y_p(x)$ is a particular integral.

COMPLIMENTARY SOLUTION OR GENERAL
SOLUTION OF A HOMOGENOUS EQUATION:

given the homogenous equation of
the form

$y'' + p(x)y' + Q(x)y = 0$

Or

$(d^2y/dx^2) + p(x)(dy/dx) + Q(x)y = 0$

if $P(x)$ and $Q(x)$ are constants

that is if we replace them with p
and q respectively then we have

$y'' + py' + qy = 0$

let $y = e^{mx}$

$y' = dy/dx =$

$d(e^{mx})/dx = me^{mx}$

and $y'' = m^2 e^{mx}$

hence

$y'' + py' + qy = 0$

can be written as

$m^2 e^{mx} + pme^{mx} + qe^{mx} = 0,$

$(m^2 + pm + q)e^{mx} = 0$

dividing through by e^{mx} we have :

$(m^2 + pm + q)(e^{mx}/e^{mx}) = 0/e^{mx}$

$(m^2 + pm + q) = 0$

using quadratic equation formula:
where $a = 1$, $b = p$, $c = q$ we then
have

$m = (-p \pm \sqrt{(p^2 - 4q)})/2$ that is

$m = (-p + \sqrt{(p^2 - 4q)})/2$ or

$m = (-p - \sqrt{(p^2 - 4q)})/2$ where

case (1.):

the roots are distinct real roots
that is where (p^2 - 4q) > 0 we
have two solution that is
y$_1$ = $_e$m$_1$x and y$_2$ = $_e$m$_2$x
and hence the general solution
where there are two distinct real
roots is
y(x) = A($_e$m$_1$x) + B($_e$m$_2$x)

Example1:

obtain the complimentary function
of the homogenous equation:
d^{2}y/dx^2 + dy/dx - 2y = 0
where d/dx is same as D then
D^{2}y + D^{1}y - 2D^{0}y = 0
= y" + y' - 2y = 0
let D^2 = m^2 ,

$D^1 = d/dx = m^1 = m$,

$D^0 = 1$

then $d^2y/dx^2 + dy/dx - 2y = 0$

becomes

$m^2 + m - 2 = 0$

since $(D^2 + D^1 - 2D^0)y = 0$

factorizing we have:

$m^2 - m + 2m - 2 = 0$

$m(m - 1) + 2(m - 1) = 0$

$(m - 1)(m + 2) = 0$

hence

$m + 2 = 0$ or $m - 1 = 0$

$m = -2$ or $m = 1$

thus $m_1 = -2$, $m_2 = 1$,

if $y = e^{mx}$

then $y_1 = e^{m_1 x} = e^{-2x}$ and $y_2 = e^{m_2 x}$

$= e^x$

the general solution then is

$y(x) = Ae^{-2x} + B e^x$

Example2:

$d^2x/dt^2 + 5dx/dt + 6x = 0$

is a homogenous equation obtain its general solution.

Solution:

$m^2 + 5m + 6 = 0$,

$m^2 + 2m + 3m + 6 = 0$

$m(m + 2) + 3(m + 2) = 0$

$(m + 3)(m + 2) = 0$

$(m + 3) = 0$ or $m + 2 = 0$,

$m = -3$

$m = -2$ that is

$m_1 = -3$ and $m_2 = -2$ or $m_1 = -2$ and $m_2 = -3$

if $y = e^{mt}$

$y_1 = e^{-2t}$

$y_2 = e^{-3t}$.

Hence the general solution is

$Ae^{-2t} + Be^{-3t}$

where case (2.):

the roots m_1 and m_2 are distinct
complex numbers (roots) that is
$p^2 - 4q < 0$.

The roots in this case can be
written as $m_1 = \alpha + Bj$ and $m_2 =$
$\alpha - Bj$. The general solution in
this case given a homogenous
equation, $ay'' + by' + cy = 0$,
is thus

let $y = e^{mx}$

where $ay'' + by' + cy = 0$ is

$am^2 + bm + c = 0$

if $m_1 = \alpha + Bj$ and $m_2 = \alpha - Bj$ *then*
the general solution where

$$y_1 = {}_e m_1 x \quad = \quad e^{(\alpha + Bj)x}$$

$$y_2 = {}_e m_2 x = e^{(\alpha - Bj)x}$$

$$y = A\ e^{(\alpha + Bj)x} + B\ e^{(\alpha - Bj)x}$$

Using indices this can also be written as

$$y = A(e^{\alpha x} . e^{Bjx}) + B(e^{\alpha x} . e^{-Bjx}) = e^{\alpha x}(Ae^{Bjx} + Be^{-Bjx}).$$

Using Euler's relations

$$e^{i\Theta} = \cos\Theta + i\sin\Theta$$

the equation $e^{\alpha x}(Ae^{jBx} + Be^{-jBx})$ can be rewritten as

$$e^{\alpha x}(A(\cos Bx + j\sin Bx) +$$

$$B(\cos Bx - j\sin Bx)) =$$

$$e^{\alpha x}(A\cos Bx + Aj\sin Bx + B\cos Bx -$$

$$Bj\sin Bx) =$$

$$e^{\alpha x}((A + B)\cos Bx + (A - B)j\sin Bx)$$

$$=$$

$$e^{\alpha x}((A + B)\cos Bx + (Aj - Bj)\sin Bx)$$

if $A + B = D_1$ and $Aj - Bj = D_2$

then the general solution is

$e^{\alpha x}(D_1 \cos Bx + D_2 \sin Bx)$.

Example:

---

solve the following homogenous
equations:

(i) $(d^2y/dt^2) + (dy/dt) + 8y = 0$

(ii) $(d^2y/dx^2) + (dy/dx) + 5y = 0$.

Solution:

(i) $(d^2y/dt^2) + (dy/dt) + 8y = 0$

then

$m^2 + m + 8 = 0$

hence a = 1, b = 1, c = 8

checking for the kind of roots we have:

$\sqrt{(b^2 - 4ac)} = \sqrt{(1 - 4(1)(8))}$

$= \sqrt{(1 - 32)} = \sqrt{(-31)}$

hence $b^2 - 4ac < 0$

thus using method of quadratic equation to solve we have:

$t = (-b \pm \sqrt{(b^2 - 4ac)})/2a$

$= (-1 \pm \sqrt{(1^2 - 32)})/2$

$= (-1 \pm \sqrt{(1 - 32)})/2$

$= (-1 \pm \sqrt{(-31)})/2 =$

$-(1/2) \pm ((\sqrt{(-31)})/2$

where $-1 = j$

we then have:

$-(1/2) \pm (\sqrt{(31j)})/2$

hence where roots are $\alpha \pm Bj$ we have

$\alpha = -(1/2)$ and

$B = (\sqrt{(31)})/2$

then the general equation is

$y =$
$e^{-(1/2)t}(D_1\cos((\sqrt{(31)})/2)t + D_2\sin((\sqrt{(31)})/2)t) = \qquad e^{-0.5t}(D_1\cos2.78t + D_2\sin2.78t).$

Solution(ii):

Given $(d^2y/dx^2) + (dy/dx) + 5y = 0$
then
$m^2 + m + 5 = 0$
checking for the kind of root we have:
$b^2 - 4ac$ where $a = 1$, $b = 1$, $c = 5$
then
$(1)^2 - 4(1)5 = 1 - 20 = 1 - 20 = -$

19 < 0

Solving thus the equation quadratically will give

x =

$(-b \pm (\sqrt{(b^2 - 4ac)}))/2a$ =

$(-1 \pm \sqrt{(1 - (4*1*5))})/2(1)$ =

$((-1 \pm \sqrt{(-19)})/2)$ =

$(-1/2) \pm (\sqrt{(-19)}/2)$

where $-1 = j$

then $-1/2 \pm (\sqrt{(-19)}/2)$ is $-(1/2)$

$\pm (\sqrt{(19j)}/2)$

thus $\alpha = -(1/2)$

$B = (\sqrt{(19)}/2)$

the general solution then is

y =

$e^{-(1/2)x}(D_1\cos((\sqrt{(19)})/2)x +$

$D_2\sin((\sqrt{(19)})/2)x)$ =

$e^{-(0.5)x}(D_1\cos(2.18x) +$

$D_2\sin(2.18x))$.

Where case (3):

where the roots m_1 and m_2 are equal
that is $b^2 - 4ac = 0$ then the
general solution of a homogenous

equation of the form
$(d^2y/dx^2) + 2a(dy/dx) + by = 0$
where a and b are constants and
(d/dx) can be written as D can be
got as below:
$(d^2y/dx^2) + 2a(dy/dx) + by = 0$ or
equivalently
$D^2y + 2aDy + by = 0 = (D^2 + 2aD + b)y = 0$
dividing through by y we have
$(D^2 + 2aD + b) = 0$.
Let $D^2 = r^2$ and $D = r$ then $r^2 + 2ar + b = 0$.
Suppose roots of the equation
above are r_1 and r_2 then. $r^2 + 2ar + b = (r - r_1)(r - r_2)$ and
$(D^2 + 2aD + b) = (D - r_1)(D - r_2)$
hence the equation $(D^2 + 2aD + b)y$

= 0 can also be written as

$(D - r_1)(D - r_2)y = 0.$

Now let $(D - r_2)y = u$

then $(D - r_1)(D - r_2)y = 0$ can be

written as $(D - r_1)u = 0.$

Solving $(D - r_1)u = 0$ by method of

separation of variables we have

$du/dx - ur_1 = 0$

$P = -r_1$

$Q = 0$

the integrating factor is

$e^{\int P dx} = e^{\int -r_1 dx} = e^{|-r_1((x^{0+1})/(0+1))} = e^{-xr_1} =$

$e^{-r_1 x}$

multiplying through by the integrating

factor then

$(du/dx) - ur_1 = 0$

becomes

$e^{-r_1 x}(du/dx) - r_1 u(e^{-r_1 x}) = 0$

then

$$u(e^{-r_1 x}) = \int 0\,dx + C$$

$$u(e^{-r_1 x}) = C$$

$$u = C(1/(e^{-r_1 x})) = C((e^{r_1 x})).$$

Substituting the above in the equation

$$(D - r_2)y = u \text{ it becomes}$$

$$(D - r_2)y = C(e^{r_1 x})$$

that is

$$(dy/dx) - r_2 y = C(e^{r_1 x}) .$$

Solving by separation of variables

$$P = r_2$$

then the integrating factor becomes

$$= e^{\int P\,dx} = e^{\int -r_2\,dx} = e^{\int -r_2 (x^{0+1}/0 + 1)}$$

$$= e^{|-r_2 x} = e^{-r_2 x}$$

multiplying $(dy/dx) - r_2 y = C(e^{r_1 x})$ by integrating factor

$$e^{-r_2 x}(dy/dx) - (r_2 y)e^{-r_2 x}$$

$$= C(e^{r_1 x}) \cdot e^{-r_2 x}$$

hence

$$e^{-r_2 x} \, y = \int (C(e^{r_1 x}) \cdot e^{-r_2 x}) \, dx$$

$$= e^{-r_2 x} \, y = C\int e^{(r_1 - r_2) x} \, dx + C_1$$

if $r_1 = r_2$ that is the roots are
equal then

$$e^{(r_1 - r_2) x} = e^0 = 1$$

hence

$$e^{-r_2 x} \, y = Cx + C$$

$$y = (Cx + C_1) \, e^{r_2 x}.$$

Example1:

obtaining the general solution of
the following homogenous
equations:

(i) $(d^2 y / dx^2) + 2(dy/dx) + y = 0$

(ii) $(d^2y/dx^2) - 4(dy/dx) + 4y = 0$.

Solution to Problem (a.):

given $(d^2y/dx^2) + 2(dy/dx) + y = 0$
we put the equation in the form:
$k^2 + 2k + 1 = 0$

Solving by factorization for the
roots we have:
$k^2 + k + k + 1 = 0$,
$k(k + 1) + 1(k + 1) = 0$,
$(k + 1)(k + 1) = 0$
hence the roots are $k = -1$ and $k =$
-1 and are thus equal. Now since k_1
$= k_2 = -1$ then the general solution
from
$y = (Cx + C_1)\,_{erx}$
is:
$y = Cx\,_{(e-1x)} + C_1\,_{(e-1x)}$.
Solution to Problem (b):
given
$(d^2y/dx^2) - 4(dy/dx) + 4y = 0$

we put the equation in the form:

$k^2 - 4k + 4 = 0$

factorizing to get the roots we have:

$k^2 - 2k - 2k + 4 = 0$

$k(k - 2) - 2(k - 2) =$

$(k - 2)(k - 2) = 0$

hence the roots are $k_1 = \quad k_2 = 2$. The general solution is thus

$Cxe^{2x} + C_1e^{2x}$.

Exercise:

1. Obtain the general solutions that is the complimentary functions of the following homogenous equations:

(a) $(d^2y/dx^2) + 7(dy/dx) + 6y = 0.$
(b) $(d^2y/dx^2) - 2(dy/dx) + y = 0.$
(c) $(d^2x/dt^2) + 5(dx/dt) + 6x = 0.$
(d) $3(d^2y/dx^2) + 4(dy/dx) + y = 0.$

Solution:

(a) $y = Ae^{-x} + Be^{-6x}$
(b) $y = Ae^x + Bxe^x$

(C) $x = Ae^{-2t} + Be^{-3t}$

(D) $Ae^{-(1/3)x} + Be^{-x}$

FINDING A PARTICULAR INTEGRAL

We stated in property 2 that the general solution of a non-homogenous equation is a sum of

the complimentary solution or the solution of the associated homogenous equation and a particular integral. We have been able to find the complimentary solution. Now we shall consider finding the particular integral. There exist many techniques for finding this particular integral however for the purpose of this book we shall use mainly a method called a method of variation of parameters. This method we can

deduce in the following manner:
given a non-homogenous equation:
$$y'' + P(x)y' + Q(x)y = R(x) \quad \ldots (1.)$$
let us assume that the general
solution of the homogenous
associate given as
$$y(x) = c_1 y_1(x) + c_2 y_2(x) \quad \ldots\ldots (2)$$
has been found. Then where we
assume c_1 to be equal to an unknown

function $v_1(x)$ and c_2 equal to an
unknown function $v_2(x)$ the
equation (2) becomes
$$y(x) = v_1 y_1 + v_2 y_2 \quad \ldots\ldots (3).$$
$$y = v_1 y_1 + v_2 y_2 \quad \ldots\ldots (3i).$$
where $v_1 y_1 = v_1(x) \cdot y_1(x)$
$v_2 y_2 = v_2(x) \cdot y_2(x)$.
Differentiating both sides of
equation 3 we have:
$$y' = v_1(y_1)' + (v_1)'y_1 + v_2(y_2)'$$
$$+ (v_2)'y_2 =$$
$$(\quad (v_1(y')_1) + (v_2(y')_2)) +$$

$(((v')_1 y_1) +$

$((v')_2 y_2)) \ldots$ equation (4).

Assuming

$(((v')_1 y_1) + ((v')_2 y_2)) = 0$

..(Equation 4i) to avoid the complication that will arise from a second differentiation of y from its y' to get y" we have:

$y' =$

$((v_1 (y')_1) + (v_2 (y')_2)) + 0$

....equation(5) differentiating y' we get

$y" = (v_1 (y")_1) + ((v')_1 (y')_1) + (v_2 (y")_2) + ((v')_2 (y')_2) \ldots (6)$

and substituting y in equ.(3i), y' in equation(5) and y" in equation(6) into the general non-homogenous equation $y" + P(x) y' + Q(x) y = R(x)$ we get

$(v_1 (y")_1) + ((v')_1 (y')_1) + (v_2 (y")_2) + ((v')_2 (y')_2)$

$+ P(x)(\quad (v_1(y')_1) + \quad (v_2(y')_2)) +$

$Q(x)(v_1y_1 + v_2y_2) = R(x) =$

$(v_1(y'')_1) + \quad ((v')_1(y')_1) +$

$(v_2(y'')_2) + ((v')_2(y')_2) +$

$P(x)(v_1(y')_1) + \quad P(x)(v_2(y')_2) +$

$Q(x)v_1y_1 + Q(x)v_2y_2 \quad = R(x)$

Collecting like terms we have,

$v_1 ((y'')_1 + (P(x)(y')_1) + Q(x)y_1) +$

$v_2((y'')_2 + \quad (P(x)(y')_2) + Q(x)y_2) +$

$((v')_1(y')_1) + ((v')_2(y')_2) = R(x)$

...equ(7).

Now since y_1 and y_2 are the solutions of the corresponding homogenous equation $y'' + P(x)y' + Q(x)y = 0$ then we can cancel out the two equations in parenthesis that is these two equations,

$v_1 ((y'')_1 + (P(x)(y')_1) + Q(x)y_1)$

and $v_2((y'')_2 + \quad (P(x)(y')_2) + Q(x)y_2)$ are equal to zero and thus we have

$0 + ((v')_1(y')_1) + ((v')_2(y')_2) =$

$R(x) = ((v')_1(y')_1) + ((v')_2(y')_2)$

$= R(x)$ ....equation(8).

Taking together equations 4(I)

that is $(((v')_1 y_1) + ((v')_2 y_2)) =$

0 and equation(8):

$((v')_1(y')_1) + ((v')_2(y')_2) = R(x)$

and solving for $(v')_2$ and $(v')_1$ we

have:

for $(v')_2$:

since $(((v')_1 y_1) + ((v')_2 y_2)) = 0$

$(((v')_1 y_1) = - ((v')_2 y_2))$

$(v')_1 = (- ((v')_2 y_2)))/y_1$

Substituting $(v')_1$ in equation (8)

we have:

$((- ((v')_2 y_2)))/y_1).(y')_1 +$

$(v')_2(y')_2 = R(x)$

$(v')_2[\{((-y_2)/y_1).(y')_1\} + (y')_2] =$

$R(x)$

making $(v')_2$ subject we have:

$(v')_2 = R(x)/[\{((-y_2)/y_1).(y')_1\} + (y')_2]$

multiplying on the right hand side the numerator and denominator by y_1 we have:

$(v')_2 =$

$((y_1)R(x))/\{[\{((-y_2)/y_1).(y')_1\} + (y')_2](y_1)\}$

$= ((y_1)R(x))/\{\{(-y_2).(y')_1\}+ ((y')_2.(y_1))\} =$

$((y_1)R(x))/\{\{((y')_2.(y_1))\}- \{(y_2).(y')_1\}\}$

where $\{\{((y')_2.(y_1))\}- \{(y_2).(y')_1\}\}$ is called Wronskian and is written as $W((y_1),(y_2))$ and hence

$(v')_2 =\{(y_1)(R(x))\}/W((y_1),(y_2))$

integrating both side we have:

$v_2 =$

$\int [\{(y_1)(R(x))\}/W((y_1),(y_2))]dx.$

For $(v')_1$:

since $(((v')_1 y_1) + ((v')_2 y_2)) = 0$

$((v')_2 y_2)) = - (((v')_1 y_1)$

thus $(v')_2 =$

$((- (((v')_1 y_1))/y_2)$

substituting $(v')_2$ in equation(8)

we have:

$((v')_1 (y')_1) +$

$((- (((v')_1 y_1))/y_2) \cdot (y')_2) = R(x)$

making $(v')_1$ subject we have:

$(v')_1 ((y')_1 - (((y_1)(y')_2)/y_2))$

$= R(x)$

dividing both sides by

$((y')_1 - (((y_1)(y')_2)/y_2))$ we get

$(v')_1 =$

$R(x)/[((y')_1 - (((y_1)(y')_2)/y_2))]$

multiplying on the right hand side

the numerator and denominator with

y_2 we get

$(v')_1 =$

$(y_2 R(x))/\{ y_2 [((y')_1 -$

$$(((y_1)(y')_2)/y_2)) \ ]\}$$

$$(v')_1 \ =$$

$$(y_2 \ R(x))/\{ \ [((y_2) \ (y')_1) -$$

$$((y_1)(y')_2) \ ]\}$$

multiplying on the right hand side the numerator and denominator by -1 we have:

$$(v')_1 \ =$$

$$(-y_2 \ R(x))/\{ \ [((y_1)(y')_2)-((y_2) \ (y')_1) \ ]\}$$

$$(v')_1 \ =$$

$$(-y_2 \ R(x))/W(y_1,y_2)$$

integrating both sides we have:

$$v_1 \ =$$

$$\int \ (((-y_2 \ R(x))/W(y_1,y_2))dx).$$

The result from equ(3i) is then:

$$y = ((y_1).(\int \ (((-y_2 \ R(x))/W(y_1,y_2))dx)) + ((y_2).(\int \ (((y_1 \ R(x))/W(y_1,y_2))dx))$$

this gives the particular
solution.

find a particular solution of
y" - 2y' + y = 2x.

the associated or corresponding
homogenous equation is.
$(d^2y/dx^2) - 2(dy/dx) + y = 0$
solving this we have: $D^2 = k^2$,
$D^1 = k$, $D^0 = 1$
hence

$k^2 - 2k + 1 = 0$,

$k^2 - k - k + 1 = 0$,

$k(k - 1) - 1(k - 1) = 0$,

$(k - 1)(k - 1) = 0$.

Hence $k - 1 = 0$ or $k - 1 = 0$ hence

$k = 1$ and thus $k_1 = k_2 = 1$

since roots are equal the general

solution is $y = Ae^x + Bxe^x$

hence from the above equation

$y_1 = e^x$

$y_2 = xe^x$

since $y_H(x) = Ay_1(x) + By_2(x)$

differentiating y_1 and y_2 we have:

$(y')_1 = e^x$

$(y')_2 = xe^x + e^x$.

hence

$w(y_1, y_2) = ((y_1)(y')_2) - ((y_2)$

$(y')_1) = e^x(xe^x + e^x) - xe^x(e^x) =$

$xe^xe^x + e^xe^x - xe^x(e^x) =$

$xe^{2x} + e^{2x} - xe^{2x} = e^{2x}$.

$w(y_1, y_2) = e^{2x}$

and thus

$V_2 = \int (((y_1 \, R(x))/W(y_1, y_2)) dx) =$

$\int (((e^x \, (R(x)=2x \,))/e^{2x}) dx)) =$

$\int ((e^x (2x))/e^{2x}) dx = \int e^x e^{-2x} (2x) dx =$

$\int e^{-x} (2x) dx = 2\int x e^{-x} dx.$

Solving by method of integration by part we have:

$u = x \, , \quad dv = e^{-x} dx$

$(du/dx) = 1; \quad \int dv = \int e^{-x} dx \, ;$

$v = (e^{-x})/-1$

thus $2\{ \, x.(\quad (e^{-x})/-1) - \quad \int ((e^-$
$^x)/-1) dx \, \} = 2\{ \, (x. \, (e^{-x})/-1) \,) + \int$
$e^{-x} dx \, \} = 2\{-x(e^{-x}) + ((e^{-x})/-1)\} =$
$2\{-x(e^{-x}) - (e^{-x})\} =$
$-2x(e^{-x}) - 2(e^{-x}).$

$V_2 = \quad -2x(e^{-x}) - 2(e^{-x})$

also $\quad V_1 = \int (((-y_2$

$R(x))/W(y_1, y_2)) dx) = \int ((-$
$x(e^x) 2x)/(e^{2x})) dx =$

$\int((-x(e^x)2x)/(e^x)^2)dx =$

$\int((-x2x)/(e^x))dx$

Or

$\int -x(e^x)(e^{-2x}).2xdx =$

$\int -x(e^{-x}).2xdx =$

$\int -(e^{-x}).2x^2dx = -2\int(e^{-x}).x^2dx =$

$-2\{(((e^{-x})/-1).x^2) + \int 2x(e^{-x})\} =$

$-2[((x^2).((e^{-x})/-1)) + 2\int x.(e^{-x})dx]$

$= -2[(-(x^2).(e^{-x})) + 2\{x.((e^{-x})/-1)$

$- \int((e^{-x})/-1)dx] =$

$-2[(-(x^2).(e^{-x})) + 2\{ -x(e^{-x}) + \int (e^{-x})dx \}] = -2[(-(x^2).(e^{-x})) + 2\{-x(e^{-x}) + ((e^{-x})/-1)\}] =$

$-2\{ -((x^2).(e^{-x})) - 2x(e^{-x}) - 2(e^{-x})\}$

$= 2(x^2.(e^{-x})) + 4x(e^{-x}) + 4(e^{-x}).$

Now since

$y = v_1y_1 + v_2y_2 = (2(x^2.(e^{-x})) + 4x(e^{-x}) + 4(e^{-x}))(e^x) + (-2x(e^{-x}) - 2(e^{-x}))x(e^x)$

since $y_1 = (e^x)$

$y_2 = x(e^x)$

thus we have

$2x^2 + 4x + 4 + (-2x^2 - 2x) =$

$2x^2 + 4x + 4 - 2x^2 - 2x =$

$4x - 2x + 4 =$

$2x + 4.$

Example 2:

find the general solution of

$(d^2x/dt^2) - 2(dx/dt) + x = e^t$

Solution:

solving for the complimentary
solution using the corresponding
homogenous equation that is
$(d^2x/dt^2) - 2(dx/dt) + x = 0$
we have
$k^2 - 2k + 1 = 0$
hence $k(k - 1) - 1(k - 1) = 0$,
$(k - 1)(k - 1) = 0$
the roots are equal and are $k_1 = 1$,
$k_2 = 1$ thus complimentary solution
is
$Ae^t + tBe^t$.
Now we solve for the particular
integral:
$y_1 = e^t$
$y_2 = te^t$
and $R(x) = e^t$
hence

$(y')_1 = e^t$
$(y')_2 = te^t + e^t$
then

$(v')_1 = ((-te^te^t)/w(y_1, y_2))$.

$v_1 = \int ((-te^te^t)/w(y_1, y_2)) \, dt$

Now since

$w(y_1, y_2) = y_1(y')_2 - y_2(y')_1 =$

$e^t \quad (te^t + e^t) - t(e^t)e^t$

$= te^te^t + e^te^t - te^te^t = te^{2t} + e^{2t} -$

$te^{2t} = e^{2t}$

thus

$v_1 = \int ((-te^te^t)/e^{2t}) \, dt \quad =$

$\int ((-te^{2t}e^{-2t}) \, dt = \int -t \, dt$

$= -\int t \, dt =$

$-| \; ((t^{1+1})/(1+1)) = - \; |((t^2)/2)$

$= -(t^2)/2$.

Solving for v_2 we have:

$v_2 = \int (y_1(R(x))/w(y_1, y_2)) \, dx$

$= \int (y_1(R(t))/w(y_1, y_2)) \, dt$

$=$

$\int ((e^te^t)/e^{2t}) \, dt \quad =$

$\int ((e^{2t})/e^{2t}) dt \qquad =$

$\int 1.dt =$

$\int t^0 .dt = (t^{0+1})/(0+1) = t$

thus $\qquad v_2 = t$

hence $y = v_1 y_1 + v_2 y_2 =$

$((-t^2)/2)e^t + t.te^t)$

$= ((-t^2)/2)e^t + t^2 e^t) =$

$(((-t^2)/2) + t^2)e^t$

$= ((-t^2 + 2t^2)/2)e^t =$

$((t^2)/2)e^t$

thus the general solution for the non-homogenous equation using property 2 is

$Ae^t + tBe^t + ((t^2)/2)e^t$

Exercise:

(1) find a particular integral for the equation.

$$d^2x/dt^2 - 3(dx/dt) + 2x = 5e^{3t}$$

(2) Find a particular integral for the equation:

$$d^2y/dx^2 + (dy/dx) + y = 1 + x.$$

(3) Find the general solution of these equations:

(a) $d^2y/dx^2 + 5(dy/dx) + 4y = 8.$

(b) $3d^2y/dx^2 + 4(dy/dx) + y = 2$

Solution:

1. $x = 2.5e^{3t}$

2. x

(3a.) $y = Ae^{-x} + Be^{-4x} + 2$

(b) $Ae^{-(1/3)x} + Be^{-x} + 6$

CHAPTER TWO

THE LAPLACE TRANSFORM

A simpler method of solving a
differential equation is by use of
a Laplace transform. This
transform is seen as a function
machine, or a device, that
transforms a differential equation
which is in the t - domain into an
algebraic equation which is in the
s-domain and is simpler to solve.
Mathematically one can define a
laplace transform as: *the result,
if it exists, of multiplying a
function of time, f(t), by e^{-st} and
integrating the result of this
multiplication between zero and
infinity.*
This is often written as:

$$L\{f(t)\} = \int_0^\infty f(t)e^{-st}dt$$

(that is integrating between 0 and ∞) where $L\{f(t)\}$ implies the laplace transform of f(t). The result of this transform is written as F(s), that is the algebraic function in the s-domain. This result once obtained can then be transformed back by an inverse operation (inverse laplace transform) of the function machine into a function in the time domain, f(t).

Example:

determine the laplace transform of
f(t) = t³.
Solution:

$$L\{f(t)\} = \int\limits_{0}^{\infty} f(t)e^{-st}dt$$

$$L\{t^3\} = \int\limits_{0}^{\infty} t^3 e^{-st}dt$$

using method of integration by
parts:

∫vdu = uv - ∫udv ,

let v = t³ and du = e⁻ˢᵗdt

differentiating v we have:

dv/dt = 3t² ,

dv = (3t²)dt

Integrating du = e⁻ˢᵗdt we have:

∫du = ∫e⁻ˢᵗdt ,

u = [(e⁻ˢᵗ)/-s]

hence,

$$\int t^3 e^{-st} dt =$$

$$\left[(e^{-st}/-s)t^3 \right]_0^\infty - \int [(e^{-st})/-s] \cdot (3t^2) dt =$$

$$\left[(e^{-st}/-s)t^3 \right]_0^\infty + (3/s) \int_0^\infty (e^{-st})t^2 dt$$

$$= 0 - 0 + (3/s) \int_0^\infty (e^{-st})t^2 dt \ ,$$

integrating further:

let v = t^2 ,

du = e^{-st}dt

hence differentiating v = t^2 we

have dv/dt = 2t,

dv = 2tdt

and integrating du = e^{-st}dt we

have:

$\int$du = $\int e^{-st}$dt ,

u = $(e^{-st})/-s$

hence

$$3/s\{\ [(e^{-st}/-s)t^2\ ]_0^\infty - \int_0^\infty (e^{-st}/-s)2t\,dt\ \}\ =$$

$$3/s\{\ [0]\ +\ 2/s \int_0^\infty e^{-st}t\,dt$$

$$=\quad 6/(s^2)\ (\ \int_0^\infty e^{-st}t\,dt\ )\ ,$$

integrating further we have let v
= t and du = (e⁻ˢᵗ)dt.
Differentiating v we get:
dv/dt = 1 ,
dv = dt.
Integrating du we get:
∫ du = ∫ (e⁻ˢᵗ)dt ,
u = [(e⁻ˢᵗ)/-s]

hence we have:

$$\{ [t.(e^{-st}/-s)]_0^\infty - \int_0^\infty (e^{-st}/-s)dt \} (6/(s^2)) =$$

$$(6/(s^2) \{ 0 + (1/s) \int_0^\infty e^{-st}dt \} =$$

$$6/(s^3) \int_0^\infty e^{-st}dt$$

$$= \quad 6/(s^3) [e^{-st}/-s]_0^\infty$$

$$= \quad 6/(s^3) | (1/s) | = 6/s^4$$

Standard Laplace transform:

there are transforms that can be used to derive transforms for a wide range of functions. A table of standard laplace transform is given below:

f(t)   L{f(t)}

1. Unit impulse(t) $\rightarrow$ 1

2. Unit step u(t)= 1 $\rightarrow$ 1/s

3. Unit ramp t $\rightarrow$ $1/s^2$

note: numbers(1) - (3) are laplace transform of integrals commonly used as input to a system.

4. t^n $\rightarrow$ $n!/(s^{n+1})$

5. e^{-at} $\rightarrow$ $1/(s + a)$

6. $1 - e^{-at}$ $\rightarrow$ $a/s(s+a)$

7. te^{-at} $\rightarrow$ $1/((s+a)^2)$

8. $e^{-at}- e^{-bt}$ $\rightarrow$ $(b-a)/((s+a)(s+b))$

9. $(1-at)e^{-at}$ $\rightarrow$ $s/((s+a)^2)$

10. $(1-(b/(b-a))e^{-at})+ (a/b-a)e^{-bt}$ $\rightarrow$
$ab/(s(s+a)(s+b))$

11. $(e^{-at}/(b-a)(c-a)) + (e^{-bt})/((c-a)(a-b)) + (e^{-at})/((a-c)(b-c))$ $\rightarrow$ $\{1/((s+a)((s+b)(s+c)))\}$

12. sinwt $\rightarrow$ $w/(s^2 + w^2)$

13. Coswt $\rightarrow$ $s/(s^2 + w^2)$

14. (e^{-at})sinwt $\rightarrow$ $w/((s+a)^2 + w^2)$

15. (e^{-at})coswt $\rightarrow$ $(s+a)/((s+a)^2 + w^2)$.

16. sinhwt $\rightarrow$ $w/(s^2 - w^2)$

17. coshwt $\rightarrow$ $s/(s^2 - w^2)$

18. $(w/\sqrt{(1-\xi^2)}).e^{-\xi\omega t}.\sin\omega\sqrt{1-\xi^2}t$ →

$\{w^2/s^2+2\xi\omega s+\omega^2\}$, where $\xi<1$.

19. $1-(1/\sqrt{(1-\xi^2)}).e^{-\xi\omega t}.\sin(\omega\sqrt{1-\xi^2}t+\phi)$ →

$\{w^2/s(s^2+2\xi ws+\omega^2)\}$, where $\cos\phi=\xi$

Properties of a laplace transform

(1a) L{f(t) + g(t)} =
L{f(t)} + L{g(t)},

that is the laplace transform of a
sum   of   time   functions   taken
together is equal to the   sum of
the   individual   separate   laplace
transforms   of   each   of   the   time
functions.

(1b) L{af(t)} = aL{f(t)} ,

this   is   so   since   2f(t)   =   f(t)   +
f(t) and L{2f(t)} = L{f(t) + f(t)}
=   L{f(t)}   +   L{f(t)}   =   2L{f(t)}.

Thus the laplace transform of a function multiplied by a constant is equal to the constant multiplying the laplace transform of the function.

determine the laplace transform of $2t + 3e^t$

$L\{2t + 3e^t\} = L\{2t\} + L\{3e^t\}$.
Solving each of the transforms individually we have:
$L\{2t\} = 2L\{t\}$ since $L\{af(t)\} =$

aL{f(t)}

hence   L{2t}   = $\int\limits_{0}^{\infty}e^{-st}2t\,dt=2\int\limits_{0}^{\infty}e^{-st}.t\,dt$

Let   v   =   t   and   du   =   e⁻ˢᵗdt

then   differentiating   v   =   t   and

integrating   du   =   e⁻ˢᵗdt   we   get:

v = t ,

dv/dt = 1

dv = dt

and

du =   e⁻ˢᵗdt

∫ du = ∫e⁻ˢᵗdt

u   =   (e⁻ˢᵗ)/-s

thus

$$2\{[t.(e^{-st}/-s)]_{0}^{\infty}-\int\limits_{0}^{\infty}(e^{-st}/-s)dt\}=2\{0+\frac{1}{s}\int\limits_{0}^{\infty}e^{-st}dt\}$$

$$=\qquad \frac{2}{s}\int\limits_{0}^{\infty}e^{-st}dt=\frac{2}{s}\left[\frac{e^{-st}}{-s}\right]_{0}^{\infty}\qquad =$$

(2/s){0 - (1/-s)} = 2/s{1/s} =

2/(s²).

For L{3e^t} = 3L{e^t} =

$$3\int_{0}^{\infty} e^{-st}.e^t dt = 3\int_{0}^{\infty} e^{(-st+t)} dt$$

$$= \qquad 3\int_{0}^{\infty} e^{(t-st)} dt = 3\int_{0}^{\infty} e^{t(1-s)} dt = 3 \left.\frac{e^{t(1-s)}}{1-s}\right|_{0}^{\infty}$$

=

3{0 - (1/1-s)}

= -3/(-s + 1) = 3/(s - 1)

hence

L{2t + 3e^t} = L{2t} + L{3e^t}

= (2/s^2)+(3/(s - 1)).

This property is true, and hence the result of the above example, because:

L{f(t) + g(t)} =

∫(f(t) + g(t))e^{-st}dt

= ∫(f(t)e^{-st}dt + g(t)e^{-st}dt) ,

$$\int_0^\infty (f(t)+g(t))e^{-st}dt = \int_0^\infty (f(t)e^{-st}dt + g(t)e^{-st}dt) =$$

$$\int_0^\infty f(t)e^{-st}dt + \int_0^\infty g(t)e^{-st}dt$$

hence L{f(t) + g(t)} =
L{f(t)} + L{g(t)}.

Example2:

determine the laplace transform of
t² + 2t + 1.

Solution:

L{ t^2 + 2t + 1} =

L{ t^2} + L{2t} + L{1}.

Using the standard laplace transforms we have:

= 2!/(s^{2+1}) + 2L{t} =

2{1!/s^{1+1}}+ 1/s =

(2*1)/s^3 = 2/(s^3) + 2(1/(s^2)) + 1/s

= 2/(s^3)+ 2/(s^2) + 1/s.

Example 3:

determine the laplace transform of sin3tcos3t.

Solution:

since 2sintcost = sin2t

thus sintcost = (sin2t)/2

hence

sin3tcos3t = (sin2(3t))/2 =

(sin6t)/2 = 1/2(sin6t) =

$1/2(6/(s^2+6^2)) = 3/((s^2)+36)$.

(2)   The   first   shift   theorem,

factor e^{-at}:

this states that if L{f(t)} = F(s)

then   L{e^{-at}f(t)}   =   F(s   +   a)

thus the replacement of s in F(s)

with s + a in F(s + a) corresponds

to multiplying a time function

f(t) with e^{-at}. This can be shown

as below:

since L{f(t)} =

$$\int_0^\infty f(t)e^{-st}dt = F(s)$$

hence

L{e⁻ᵃᵗf(t)} =

$$\int_0^\infty f(t)e^{-st}e^{-at}\,dt =$$

$$\int_0^\infty f(t)e^{-(s+a)t}\,dt$$

= F(s + a).

Example1:

determine the laplace transform of
e³ᵗcost.

Solution:

where we have:
L{e³ᵗcost}

then a = -3 since e^{3t} . Where f(t)
= cost using standard laplace
transform,

F(s) = s/(s² + 1)

hence F(s + (-3)) =

(s + (-3))/((s + (-3))² + 1)

= (s - 3)/((s - 3)²+ 1)

= (s - 3)/(s² - 6s + 9 + 1)

= (s - 3)/(s² - 6s + 10).

Example2:

determine the laplace transform of
$t^3 e^{-3t}$.

Solution:

from the equation $t^3 e^{-3t}$

a = 3.

hence   using   standard   laplace
transforms we have

$f(t) = t^3$ and hence

$L\{t^3\} =$

$3!/s^{3+1} = (1.2.3)/(s^4)$

$= 6/(s^4)$

$= F(s)$

$\therefore F(s + a) = F(s + 3)$

$= 6/(s + 3)^4$.

(3) *Second   shift   theorem,   time
shifting:*

this states that if a signal/function is delayed by a period or time T then its laplace transform is multiplied by a factor e^{-sT}. A signal u(t) that is delayed by a period T is written as u(t - T) where T is the delay. This implies that since F(s) is the laplace transform of f(t) and f(t) = 1*f(t) where u(t) = 1 then L{f(t - T)u(t - T)} = e^{-sT}F(s).

Example:

use the second shift theorem to determine the laplace transform of the following functions:

(a) a unit step function which starts at t = 5s.

(b) the function described by 3(t - 10)u(t - 10).

(a) f(t) = 1 and t = 5s hence we integrate f(t) = 1 between 5 and infinity or starting from 5 to infinity(∞) that is

$$\int_5^\infty 1e^{-st}\,dt = \left. e^{-st}\middle/{-s}\right|_5^\infty =$$

$$0 + (e^{-s5}/s) = \left(\frac{1}{s}\right)\left(\frac{1}{e^{5s}}\right),$$

$$= \frac{1}{se^{5s}}$$

(b) Given 3(t - 10)u(t - 10) then we have f(t) = 3t. Since we have

$3(t - 10)u(t - 10)$ hence $L\{3t\} = 3/(s^2)$. Thus using the 2nd shift theorem we have since the delay T $= 10$ from $e^{-sT}(F(s))$ the result: $(e^{-s10})(3/(s^2)) = (3(e^{-10s}))/(s^2)$.

THE LAPLACE TRANSFORM OF DERIVATIVES

the laplace transform of the first derivative of $f(t)$ that is $df(t)/dt$ is written as $L\{df(t)/dt\}$. This can be deduced as below:

$L\{df(t)/dt\} =$

$$\int_0^\infty e^{-st}\left(\frac{df(t)}{dt} \right) dt$$

Let $dv = df(t)dt/dt$ and $u = (e^{-st})$ then $\int dv = \int (df(t)/dt)dt$

= f(t) = v

du/dt = -s(e^(-st))

du = -s(e^(-st))dt.

$$\left[e^{-st}f(t)\right]_0^\circ - \int f(t) - se^{-st}dt$$

$$=$$

$$\left[e^{-st}f(t)\right]_0^\circ + s\int_0^\infty f(t)e^{-st}dt$$

= -f(0) + sF(s)

hence L{df(t)/dt} = sF(s) - f(0).

The laplace transform of the second derivative is:

$$L\{d^2f(t)/dt^2\} = \int_0^\infty \left(d^2f(t)\middle/dt^2\right)e^{-st}dt =$$

$$\int_0^\infty \left(d^2f(t)\middle/dt^2\right)e^{-st}dt$$

let dv = {d²f(t)/dt²} ; u = (e^(-st))

Integrating dv and differentiating u we have:

∫ dv = ∫ {d²f(t)/dt²}dt,

v = df(t)/dt ,

u =   (e⁻ˢᵗ),

du/dt = -s(e⁻ˢᵗ),

du = -s(e⁻ˢᵗ)dt.

Thus,

$$\left[e^{-st}\left(\frac{df(t)}{dt} \right) \right]_0^\infty - \int_0^\infty \left(\frac{df(t)}{dt} \right) - se^{-st}\,dt$$

$$=$$

$$-\left(\frac{df(0)}{dt} \right) + s\int_0^\infty \left(\frac{df(t)}{dt} \right) e^{-st}\,dt$$

$$=$$

(-df(0)/dt) + s{sF(s) - f(0)} =

(-df(0)/dt) + s²F(s) - sf(0) =

s²F(s) - sf(0)- df(0)/dt.

Likewise the 3ʳᵈ derivative by the same method is

L{d³f(t)/dt³} =

s³F(s)-s²f(0)-s(df(0)/dt)-

(d²f(0)/dt²).

Examples:

(1) the laplace transform of y(t) is Y(s), y(0) = 3, y'(0) = 1. Find the laplace transform of the following expression:

(a) y" + 2y' + 3y.

$L\{y''\} = s^2Y(s) - sY(0) - Y'(0)$,

$L\{y'\} = sY(s) - Y(0)$,

$L\{y\} = Y(s)$.

$\therefore L\{y'' + 2y' + 3y\} =$

$L\{y''\} + 2L\{y'\} + 3L\{y\} =$

$\{s^2Y(s) - sY(0) - Y'(0)\} + 2\{sY(s) - Y(0)\} + 3Y(s)$

since Y(0) = 3 and Y'(0) = 1 then

$L\{y'' + 2y' + 3y\} =$

$\{s^2Y(s) - s3 - 1\} + 2\{sY(s) - 3\}$

$+ 3Y(s)$

$= s^2Y(s) - 3s - 1 + 2sY(s) - 6 +$

$3Y(s)$

$=$

$s^2Y(s) + 2sY(s) + 3Y(s) - 3s - 7 =$

$Y(s)\{ s^2 + 2s + 3\} - 3s - 7.$

(2)Given the laplace transform of f(t) is F(s). Where f(0) = 2 and f'(0) = 3, find the laplace transform of 3f' - 2f.

Solution:

$L\{f'\} = sF(s) - f(0),$

$L\{f\} = F(s).$

$\therefore L\{3f' - 2f\} = 3L\{f'\} - 2L\{f\} =$

$3\{ sF(s) - f(0) \} - 2F(s)$

$= 3sF(s) - 3f(0) - 2F(s).$

Since $f(0) = 2$, $f'(0) = 3$ then

$L\{3f' - 2f\} =$

$3sF(s) - 3(2) - 2F(s) =$

$F(s)\{3s - 2\} - 6.$

(b) f''.

Solution:

$L\{f''\} = s^2F(s) - sf(0) - f'(0).$

Since $f'(0) = 3$ and $f(0) = 2$ then

$L\{f''\} = s^2F(s) - 2s - 3.$

THE LAPLACE TRANSFORM OF INTEGRAL: consider the laplace transform of the integral of a function that is:

$$L\{\int_0^t (f(t))dt\}$$

let g(t) = $\int_0^t (f(t))dt$, then

dg(t)/dt = f(t)

since dg(t) = f(t)dt

= ∫dg(t) = ∫f(t)dt

= g(t) = ∫f(t)dt.

Now consider the laplace transform of

dg(t)/dt that is

L{dg(t)/dt}.

This since is

L{dg(t)/dt} = sG(s) - g(0)

and since G(s) = L{g(t)}

then L{dg(t)/dt} = L{f(t)}

= sL{g(t)} - g(0)

since

f(t) = dg(t)/dt.

Where g(0) = 0 then

L{f(t)} = sL{g(t)} - 0

= L{f(t)} = sL{g(t)}

dividing both sides by s :

L{f(t)}(1/s) = (1/s).sL{g(t)}

(1/s). L{f(t)} = L{g(t)}

$$(1/s) F(s) = L\{\int_0^t (f(t))dt\}$$

since L{f(t)} = F(s).

Example:

determine the laplace transform of

$$\int_0^t e^{-2t}\, dt.$$

Solution:

$$L\{\int_0^t (f(t))dt\} = (1/s)F(s)$$

where $f(t) = e^{-2t}$

$F(s) = L\{e^{-2t}\} = 1/(s + 2)$

$$\therefore L\{\int_0^t e^{-2t}dt\} = (1/s)(1/(s + 2))$$

$$= 1/(s(s + 2)).$$

Exercise

1. Find the Laplace transform of the following equations:

(a) $2 - t^2 + 2t^4$

(b) $2sin4t + 11 - t$

2. The laplace transform of y(t) is Y(s), y(0)=3, y'(0)= 1. Find the Laplace transform of the following expression:

(a) $3(d^2y/dt^2) + 6(dy/dt) + 8y$

Solution

1. (a) $(2/s) - (2/s^3) + (48/s^5)$

(b) $(8/(s^2+16)) + (11/s) - (1/s^2)$

2. $(3s^2 + 6s + 8)Y - 9s - 21$

CHAPTER THREE

THE INVERSE LAPLACE TRANSFORM:
this involves the transformation
of a laplace transform that is a
function of s, F(s), into a
function in time domain that is
f(t). Take the Laplace transform
F(s) for an example. Its inverse
laplace transform is written as
$L^{-1}\{F(s)\}$ and this is equal to
f(t).

The inverse Laplace transform has
basic properties. These properties
together with the standard Laplace
transforms can be used to obtain a
wide range of transforms. However
Laplace transforms, F(s), that
occurs as a ratio of polynomials
cannot often be easily expressed
using standard Laplace transforms.
Hence we use partial fractions to
split the polynomials into simpler
fractions that can be identified

with a standard Laplace transform.

Basic properties of inverse laplace transforms:

(1) Additive property:

(a) $L^{-1}\{aF(s)\} = aL^{-1}\{F(s)\}$

where a is a constant and $L^{-1}\{F(s)\}$ is the inverse laplace transform of f(t).

(b) the sum of two separate inverse laplace transforms is equal to the inverse laplace transform of the sum taken together. That is

$L^{-1}\{F(s) + G(s)\} = L^{-1}\{F(s)\} + L^{-1}\{G(s)\}$.

(2) First shift theorem:

this can be written as:

$L^{-1}\{F(s - a)\} = e^{at}f(t)$.

(3) Second shift theorem:

this in inverse form can be written as:

$L^{-1}\{e^{-sT}F(s)\} = f(t - T)u(t - T)$.

Solving for the inverse using the

theorem involves:

(1) remove e^{-sT} where it is in the numerator of the transform.

(2) find the inverse of what remains after removing e^{-sT}.

(3) substitute in the result obtained t - T for t that is replace t with t - T.

Example1:

determine the inverse laplace transforms of e^{-2s}/s^2.

Solution:

$e^{-2s}/s^2 = e^{-2s}(1/s^2)$

here: $F(s) = (1/s^2)$ and $T = 2$

thus $L^{-1}\{F(s)\} = L^{-1}\{(1/s^2)\} = t$

hence by 2nd shift theorem the inverse transform is

$(t - 2)u(t - 2)$

since $f(t) = t$ and if $f(t) = t$ then $f(t - 2) = t - 2$.

Example2:

determine the inverse laplace transforms of:

$e^{-3s}/(s + 2)^2$.

Solution:

Given $e^{-3s}/(s + 2)^2$

$F(s) = 1/(s + 2)^2$ and $T = 3$

thus

$L^{-1}\{1/(s + 2)^2\} = te^{-2t}$

and hence

$L^{-1}\{e^{-3s}/(s + 2)^2\} =$

$(t - 3)e^{-2(t - 3)}u(t - 3)$.

Example4:

determine by the use of partial fraction, the inverse laplace transform of

$(3s + 1)/(s^2 - s - 6)$.

Solution:

where: $(s^2 - s - 6)$

factorizing we have:

$(s^2 + 2s - 3s - 6) =$

$s(s + 2) - 3(s + 2) =$

$(s - 3)(s + 2)$

hence :

$(3s + 1)/(s^2 - s - 6) =$

$3s + 1/(s - 3)(s + 2)$

solving the above using partial fraction method we have:

$3s + 1/(s + 2)(s - 3) =$

$(A/(s - 3)) + (B/(s + 2)) =$

$(A(s + 2) +$

$B(s - 3))/((s - 3)(s + 2)) =$

$(As + 2A + Bs - 3B)/(s-3)(s+2)$

Equating coefficients of s:

$As + Bs = 3s$

$(A + B)s = 3s$

$A + B = 3 \ldots (1)$

$2A - 3B = 1 \ldots\ldots (2)$

Multiplying equ(1.) by 3 and solving simultaneously we have:

$3A + 3B = 9 \ldots\ldots (3)$

and hence equ.(3) + equ.(2) will be:

5A = 10

5A/5 = 10/5

A = 2

From equ.(1):

A + B = 3

thus B = 3 - A

where A = 2

then B = 3 - A

becomes B = 3 - 2 = 1

hence $((3s) + 1)/(s^2 - s - 6) =$

$(2/(s - 3)) + (1/(s + 2))$

thus $L^{-1}\{(3s + 1)/(s^2 - s - 6)\} =$

$L^{-1}\{(2/(s - 3)) + (1/(s + 2))\} =$

$L^{-1}\{(2/(s - 3))\} + L^{-1}\{(1/(s + 2))\}$

$=$

$2L^{-1}\{(1/(s - 3))\} + L^{-1}\{(1/(s + 2))\}$

$= 2e^{3t} + e^{-2t}$.

Example 5:

determine   by   use   of   partial
fractions   the   inverse   laplace
transform of

(s - 4)/((s + 1)(s² + 4))

simplify the equation by use of
partial fraction:

(s - 4)/((s + 1)(s² + 4)) =

(A/(s + 1)) + ((Bs + C)/(s² + 4)) =

(A(s² + 4) +

(Bs + C)(s + 1))/((s + 1)(s² + 4))

$$= \frac{A(s^2+4)+(Bs+C)(s+1)}{(s+1)(s^2+4)} =$$

(A(s²) + 4A + B(s²) + Bs + Cs +

C)/((s + 1)(s² + 4)) =

$$\frac{As^2 + 4A + Bs^2 + Bs + Cs + C)}{(s+1)(s^2+4)}$$

hence equating on the RHS and LHS
coefficients and constants we
have:

A(s²) + B(s²) = 0

(A + B)s² = 0

A + B = 0 .... (1)

A = -B

also

Bs + Cs = s

(B + C)s = s

B + C = 1 ..... (2)

also

4A + C = -4 ..... (3)

since A = -B then

equation(2) becomes

-A + C = 1 .... (4.)

solving (3) and (4) simultaneously

we have:

equation(3) - equation(4) :

5A = -5

thus 5A/5 = -5/5

A = -1

hence from equation(1)

A + B = 0

B = -A

= -(-1) = 1

from equation 4:

-A + C = 1

-(-1) + C = 1

1 + C = 1

C = 1 - 1 = 0

thus

$(A/(s + 1)) + ((Bs + C)/(s^2 + 4)) =$

$-1/(s + 1) + ((1(s) + 0)/(s^2 + 4))$

$= (-1/(s + 1)) + (s/(s^2 + 4))$

the inverse is thus

$L^{-1}\{ (-1/s + 1) + (s/(s^2 + 4))\} =$

$L^{-1}\{ -1/(s + 1)\} + L^{-1}\{ s/ (s^2 + 4)\}$

$= -1e^{-t} + cos2t.$

Example 6:

determine   the   inverse   laplace
transforms of
$5/((s - 2)^2 + 25)$

$5/ ((s - 2)^2 + 25) =$
$5/((s - 2)^2 + 5^2)$
$= \quad \sin5t.e^{2t}$.

INITIAL AND FINAL VALUE THEOREMS:
the initial value of a function
that is the value of a function at
t = 0 and the final value, that is
the function's value at t = ∞, can
be   determined   from   a   laplace
transform without the need for the
use of inverse laplace transform.

This is achievable through the initial and final value theorem.

THE INITIAL VALUE THEOREM:

this states that

$$\lim_{t->0} f(t) = \lim_{s->\infty} sF(s)$$

This can be proven as below:

take laplace transform of f(t),

$$L\{f(t)\} = \int_0^\infty f(t)e^{-st}dt$$

and the laplace transform of df(t)/dt, L{(df(t)/dt)} =

$$\int_0^\infty e^{-st}\left(\frac{df(t)}{dt}\right)dt =$$

sF(s) - f(0)

then taking limits as s -> ∞ for L{df(t)/dt} that is

$$\lim_{s->\infty}\int_0^\infty e^{-st}\left(\frac{df(t)}{dt}\right)dt =$$

$$\lim_{s->\infty}\left[sF(s) - f(0)\right]$$

we will have

$$\lim_{s\to\infty}\left[sF(s)-f(0)\right] = 0$$

$$= \lim_{s\to\infty}\left[sF(s)\right]-\lim_{s\to\infty}f(0) = 0,$$

$$\lim_{s\to\infty}\left[sF(s)\right]=\lim_{s\to\infty}f(0)$$

since as s -> ∞, e⁻ˢᵗ -> 0 and

(df(t)/dt) -> 0

now since $\lim_{s\to\infty}f(0)$ = f(0)

as limit of a constant is the constant and f(0) is the value of a function at t = 0 then

$$\lim_{t\to 0}f(t)=\lim_{s\to\infty}sF(s)$$

Example1:

determine the initial values of the functions giving the following laplace transforms:

(a) 1/(s² + 1)   (b) s/(s² + 1).

(a) $1/(s^2 + 1)$

since $\lim_{t->0} f(t) = \lim_{s->\infty} sF(s)$

then where F(s) = $1/(s^2 + 1)$

we have:

$$\lim_{s->\infty} s\left(\frac{1}{s^2+1}\right) = \lim_{s->\infty}\left(\frac{s}{s^2+1}\right)$$

dividing both numerator and

denominator by s we get:

$$\lim_{s->\infty}\left(\frac{\frac{s}{s}}{\frac{s^2}{s}+\frac{1}{s}}\right) = \lim_{s->\infty}\left(\frac{1}{s+\frac{1}{s}}\right)$$

$$= \lim_{s->\infty}\left(\frac{1}{s+\frac{1}{s}}\right)$$

$$= \left(\frac{1}{\infty+\frac{1}{\infty}}\right) = \left(\frac{1}{\infty+0}\right) = 0.$$

(b) F(s) = $s/(s^2 + 1)$ hence

$$\lim_{s->\infty} sF(s) = \lim_{s->\infty} s\left(\frac{s}{s^2+1}\right) = \lim_{s->\infty}\left(\frac{s^2}{s^2+1}\right)$$

dividing numerator and denominator

by s^2 we have:

$$\lim_{s->\infty}\left(\frac{\frac{s^2}{s^2}}{\frac{s^2}{s^2}+\frac{1}{s^2}}\right)$$

$$=\lim_{s->\infty}\left(\frac{1}{1+\frac{1}{s^2}}\right)$$

hence as s -> ∞

$$=\lim_{s->\infty}\left(\frac{1}{1+\frac{1}{s^2}}\right)=\frac{1}{1+\frac{1}{\infty^2}}$$

$$=\left(\frac{1}{1+0}\right)=1$$

FINAL VALUE THEOREM:

given that

L{df(t)/dt} =

$$L\left\{\frac{df(t)}{dt}\right\}$$

=

$$\int_0^\infty e^{-st}\cdot\left(\frac{df(t)}{dt}\right)dt=$$

{sF(s) - f(0)}

then the limit of L{df(t)/dt} as s

-> 0

is written as:

$$\lim_{s \to 0} \left\{ \int_0^\infty e^{-st} \cdot \left(\frac{df(t)}{dt} \right) dt \right\} =$$

Now as s -> 0, e⁻ˢᵗ -> 0

hence we have:

$$\int_0^\infty \cdot \left(\frac{df(t)}{dt} \right) dt$$

$$=$$

$$\lim_{t \to \infty} \int_0^t \left(\frac{df(t)}{dt} \right) dt = \lim_{t \to \infty} [f(t) - f(0)].$$

Now since

$$\int_0^\infty e^{-st} \cdot \left(\frac{df(t)}{dt} \right) dt = sF(s) - f(0)$$

hence

$$\lim_{s \to 0}\{sF(s) - f(0)\}$$

$$= \lim_{t \to \infty}[f(t) - f(0)] =$$

$$\lim_{s \to 0} sF(s) - \lim_{s \to 0} f(0)$$

$$=$$

$$\lim_{t \to \infty} f(t) - \lim_{t \to \infty} f(0)$$

$$=$$

$$\lim_{s \to 0} sF(s) = \lim_{t \to \infty} f(t)$$

since $\lim_{s \to 0}\{f(0)\} = 0$

and $\lim_{t \to \infty}\{f(0)\} = 0$

hence

$$\lim_{t \to \infty} f(t) = \lim_{s \to 0} sF(s)$$

the above is termed the final value theorem.

Examples:

determine the final values of the functions having the following laplace transforms:

(a) 2/s   (b) 1/(s + 5).

133

Solution:

(a) since $\lim_{t \to \infty}\{f(t)\} = \lim_{s \to 0} sF(s)$

and F(s) = 2/s

hence

$$\lim_{s \to 0} s\left(\frac{2}{s}\right) = \lim_{s \to 0}\left(\frac{2s}{s}\right)$$

$$= \lim_{s \to 0} 2 = 2.$$

(b) F(s) = 1/(s + 5) hence

$$\lim_{s \to 0} sF(s) = \lim_{s \to 0} s\left(\frac{1}{s+5}\right)$$

$$= \lim_{s \to 0}\left(\frac{s}{s+5}\right)$$

hence the value of $\left(\frac{s}{s+5}\right)$ as s->0

(that is evaluating $\left(\frac{s}{s+5}\right)$ at the point

s is approximately equal to 0, that is as it would

result when s = 0)

$$= 0/(0 + 5) =$$

$$\left(\frac{0}{5}\right) = 0.$$

Exercise

1. Find the inverse laplace transforms of the following functions: (a) $6s/(s^2-8)$ (b) $(3s-7)/(s^2+9)$

2. Express the following as partial fractions and hence find the inverse Laplace transforms: (a) $(7s+3)/(s(s+3)(s+4))$

Solution

1(a) $6\cosh\sqrt{8}t$

(b) $3\cos3t-(7/3)\sin3t$

2(a) $(¼) + 6e^{-3t} - (25/4)e^{-4t}$

CHAPTER FOUR

SOLVING DIFFERENTIAL EQUATIONS:

Laplace transform provides a way
of solving differential equations.
Examples of this procedure are
provided below:

Example 1:

solve the differential equations:

$$\left(\frac{d^2x}{dt^2}\right) + x = 3$$

where x = 0 and dx/dt = 1 when t =
0.

given the equation $\left(\dfrac{d^2x}{dt^2}\right) + x = 3$

we follow the procedure below to solve the problem:

(a) change each term in the equation into its corresponding laplace transform:

$\{s^2X(s) - sX(0) - X'(0)\} + X(s) = 3/s$.

(b) substitute the initial condition values in the equation and solve algebraically:

$\{s^2X(s) - s(0) - 1\} + X(s) = 3/s$

$s^2X(s) + X(s) = (3/s) + 1$

$(s^2 + 1)X(s) = 3/s + 1/1$

$= (3 + s)/s$

$X(s) = ((3 + s)/s)(1/(s^2 + 1))$

(c) use inverse transform to change each term to its inverse or function of t:

$X(s) = ((3 + s)/s(s^2 + 1))$.

Resolving by partial fraction we have:

$((3 + s)/s(s^2 + 1)) =$

$(A/s) + ((Bs + C)/(s^2 + 1)) =$

$$\frac{A}{s} + \left(\frac{Bs + C}{(s^2 + 1)} \right) =$$

$(A(s^2 + 1) + (Bs + C)s)/(s(s^2 + 1))$

$=$

$(A(s^2) + A + B(s^2) + Cs)/(s(s^2 + 1))$

$$= \left(\frac{As^2 + A + Bs^2 + Cs}{s(s^2 + 1)} \right)$$

Equating coefficients we have:

$A(s^2) + B(s^2) = 0 \ldots (1)$

$Cs = s =$

$C = 1$

$A = 3$

$(A + B)(s^2) = 0$

$A + B = 0$

$B = -A = -(3) = -3$

hence $\{(s + 3)/s(s^2 + 1)\} =$

$(3/s) + \{(-3s + 1)/(s^2 + 1)\}$.

Resolving further the equation in bracket we have:

$\{(-3s + 1)/(s^2 + 1)\} = \{A/(s^2 + 1)\}$ $+ \{Bs/(s^2 + 1)\} = (A + Bs)/(s^2 + 1)$

equating coefficients:

$Bs = -3s$

$B = -3$

$A = 1$

hence $[(-3s + 1)/(s^2 + 1)] =$ $(1/(s^2 + 1)) + (-3s/(s^2 + 1))$

thus $\{(3 + s)/s(s^2 + 1)\} =$ $(3/s) + (1/(s^2 + 1)) -$ $(3s/(s^2 + 1))$

Now taking the inverse laplace transform for each in the above equation we have:

$3L^{-1}\{(1/s)\} + L^{-1}\{(1/(s^2 + 1))\} +$ $3L^{-1}\{(s/(s^2 + 1))\}$ $=$

$3 + \text{sint} - 3\text{cost}$.

solve the following differential
equations:

$$\frac{d^2x}{dt^2} + 4x = 1$$

with x = 0 and dx/dt = 0 when t =
0.

Solution:

taking laplace on both sides of
the equation we have:

L{(d²x/dt²) + 4x} = L{1}

s²X(s) - sX(0) - X'(0) + 4X(s) =
1/s

at t = 0, X = 0 and dx/dt = 0

hence

s²X(s) - 0 - 0 + 4X(s) = 1/s

s²X(s) + 4X(s) = 1/s

X(s)(s² + 4) = 1/s

X(s) = (1/s).(1/(s² + 4)) =
(1/[s(s² + 4)]).

Solving using partial fraction:

$(1/s(s^2 + 4)) =$

$(A/s) + (Bs/(s^2 + 4)) =$

$(A(s^2 + 4) + B(s^2))/(s(s^2 + 4)).$

equating coefficients of

numerator:

$As^2 + 4A + Bs^2 = 1$

$As^2 + Bs^{2.} = 0$

$(A + B)s^2 = 0$

$A + B = 0$

and

$4A = 1$

thus $A = 1/4$

$A + B = 0$

$B = -A = -(1/4)$

hence

$(1/[s(s^2 + 4)]) = ((1/4)/s) + [((-1/4)s)/(s^2 + 4)] \quad =$

$((1/4)/s) - (((1/4)s)/(s^2 + 4))$

now taking the inverse laplace

transform:

$$L^{-1}\left\{\frac{\frac{1}{4}}{s} - \frac{\frac{1}{4}s}{s^2+4}\right\} =$$

L⁻¹{((1/4)/s) - (((1/4)s)/(s² +

4))} =

$$L^{-1}\left\{\left(\frac{1}{4}\right)\frac{1}{s}\right\} - L^{-1}\left\{\frac{1}{4}\left(\frac{s}{s^2+4}\right)\right\}$$

=(1/4)L⁻¹{(1/s)}-(1/4)L⁻¹{(s/(s² +

4))} = 1/4(1) - (1/4)cos2t

=   1/4 - (1/4)cos2t.

Exercises:

use laplace transforms to solve:

(a) 3(dx/dt) + 4x = 2,

x(0) = 2.

(b) (dx/dt)+ 5x = 2

with x = 0 when t = 0.

(c)2(dx/dt) + x = 4e²ᵗ

with x = 0 when t = 0.

(a) $x = (3/2) e^{-(4t/3)} + (1/2)$

(b) $x = (2/5) - (2/5) e^{-5t}$

(c) $x = 3e^{2t} - 3e^{-(t/2)}$

CHAPTER FIVE

TRANSFER FUNCTION

Transfer function is a ratio of output of a system to the input of a system when all initial conditions before we apply the input is zero. This is denoted by

$G(s) = Y(s)/X(s)$

where $G(s)$ = Gain,

$Y(s)$ = output

and $X(s)$ = input.

This is a more convenient form of describing the relationship between the output and input of a system than using the differential equations. Hence a system whose input and output relationship have been described by a differential equation is transformed using a Laplace transform and then described using a transfer

function. A transfer function is
thus a useful tool in constructing
the mathematical model of a
system.

Example:

given the equation of a system as
$(d^2x/dt^2) + 3(dy/dx) - 4y = x$
x is input, y is output; and y = 0
and dy/dx = 0 when x = 0
determine the transfer function.

Solution:

change the differential equation
using laplace transform to an s-
domain function:

$s^2Y(s) - sY(0) - Y'(0) + 3(sY(s) - Y(0)) - 4Y(s) = X(s)$

since $Y(x=0) = 0$, $dy/dx = 0$

hence $s^2Y(s) + 3sY(s) - 4Y(s) = X(s)$

separate Y(s):

$Y(s)[s^2 + 3s - 4] = X(s)$

dividing both sides by Y(s):

$[s^2 + 3s - 4] = X(s)/Y(s)$

$Y(s)/X(s) = [1/(s^2 + 3s - 4)]$

thus,

$G(s) = Y(s)/X(s) =$

$[1/(s^2 + 3s - 4)]$

$G(s)$ = transfer function.

Systems in series:

consider a system that has two
subsystems in series. If the input
to the first system is X(s) and
the output $Y_1(s)$ and the input to

the second system $Y_1(s)$, output of
the first, and its output $Y(s)$
then the transfer function $(G(s))$
of such a system with two
subsystems can be expressed as:
$G_2(s) = Y(s)/Y_1(s)$
where $Y_1(s)/X(s) = G_1(s)$
$Y_1(s) = G_1(s).X(s)$
$G_2(s) = [Y(s)/(G_1(s).X(s))]$
$Y(s) = G_2(s).G_1(s).X(s)$
$G_2(s).G_1(s) = Y(s)/X(s)$
$Y(s)/X(s) = $ overall $G(s)$.
Thus the total transfer function
of a system composed of several
subsystems in series is the
product of the transfer function
of each subsystem that make up the
system.

determine the overall transfer
function of a system consisting of
two elements in series with one
element having a transfer function
of 1/(s - 2) and the other of 1/(s
+ 4).

G(s)[overall] = G_1(s).G_2(s) =
(1/(s - 2))*(1/(s + 4)) =
1/((s - 2)(s + 4)).

Systems with negative feedback:
a transfer function can be
represented as a block diagram. A
simple example can be shown below:

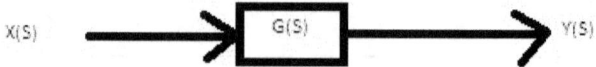

X(s) = input

Y(s) = output

G(s) = transfer function, or a
system, shown as a box. This
receives the input signal and
produces an output, Y(s).
A second example is a system with
a negative feedback loop. This is
shown below:

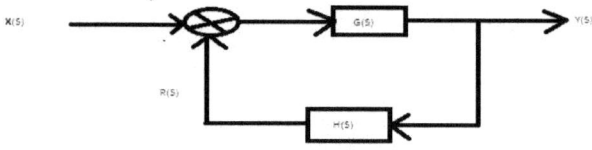

Considering the system one can
observe the following:
(1) the input to the system H(s)

is the output of the system $G(s)$,
that is $Y(s)$. This is an output of
a system fedback as an input to
another system.

(2) the output of the system $H(s)$
that is $Y(s)*H(s)$ since

$H(s) = (R(s)/Y(s))$ and hence

$R(s) = H(s)*Y(s)$

where $H(s)$ = feedback system, $R(s)$
= feedback signal and $X(s)$ is the
input to the system $G(s)$ ($G(s)$ is
the forward path system).

(3) $R(s)$ is subtracted as can be
shown by the negative sign from
$X(s)$, the input to $G(s)$, and the
result is fed back as input to
$G(s)$. This result is called the
error signal. Thus error signal =
$X(s) - H(s)Y(s)$ and hence

$G(s) = Y(s)/[X(s) - H(s)Y(s)]$

$G(s)[X(s) - H(s)Y(s)] = Y(s) =$

$X(s)G(s) - H(s)Y(s)G(s) = Y(s) =$

X(s)G(s) = H(s)Y(s)G(s) + Y(s)

G(s)X(s) = Y(s)[H(s)G(s) + 1]

dividing both sides thro by

[H(s)G(s) + 1] we have:

(G(s)X(s)/[H(s)G(s) + 1])= Y(s).

dividing further by X(s) on both

sides we have:

[G(s)X(s)/(H(s)G(s) + 1)]*(1/X(s))

= Y(s)/X(s)

(Y(s)/X(s)) = G(s)/(G(s)H(s) + 1)

hence the overall transfer

function for a feedback system is

G(s)/((G(s)H(s)) + 1) = $\dfrac{G(s)}{G(s)H(s)+1}$.

Example:

determine the overall transfer

function of a control system with

the transfer function for its

negative feedback loop as 8 and
the forward path transfer function
as 4/(s + 2).

G(s) = (4/(s + 2)), H(s) = 8
overall transfer function =
((4/(s + 2))/(((4/(s + 2))* 8) +
1)) =
(4/(s + 2))/((32/(s + 2)) + 1) =
(4/(s + 2))/((32 + s + 2)/(s + 2))
= (4/(s + 2))/((s + 34)/(s + 2)) =
(4/(s + 2))*((s + 2)/(s + 34)) =
4/(s + 34).

Obtaining the outputs of systems:
obtaining the output of a system
given as an s-function as a
function of time:

we achieve this by first changing
the given differential equation
into an s-function algebraic
equation with the given initial
conditions in mind, obtaining from
the obtained s-function the output
of the s-function and then
changing the output s-function
into a time function using
standard Laplace transform.
The output of a first order
system:
a first order system has a
differential equation of the form:
τ (dy/dt) + y = kx
changing this to an s-function we
have:
τ sY(s) + Y(s) = KX(s) ,
Y(s)(sτ + 1) = KX(s) ,
(Y(s)/X(s)) = (k/(τ s + 1)) = G(s).
Y(s) = ((k/(τ s + 1))*X(s)) =
KX(s)/(τ s + 1)

where Y(s) is output, X(s) = 1/s when the first order system is subjected to a unit step input. It is 1, that is X(s) = 1, when the first order is subjected to a unit impulse input. Hence for Y(s) =

$$((k/(\tau s + 1)) * X(s)) = \left(\frac{k}{\tau s + 1}\right) X(s)$$

for a unit step input we will have:

Y(s) = (k/(τ s + 1)) * (1/s) =
k/(s(τ s + 1)) =
(k/(τ s² + s)) * 1.

Dividing the numerator and denominator by τ we have:
(k/τ)/((τ s²/τ) + (s/τ)) =
((1/τ) * k)/(s² + (s/τ)) =

$$\left(\frac{\left(\frac{1}{\tau}\right)k}{s\left(s + \frac{1}{\tau}\right)}\right) =$$
((1/

τ) * k)/(s(s + (1/τ))).

Using the standard Laplace

transform we have:

$= (k(1/\tau))/(s(s + (1/\tau))) =$

$k(1-e^{-\frac{1}{\tau}t}) =$

$k(1-e^{-\frac{t}{\tau}})$

for a unit impulse input we have:

X(s) = 1.

Hence Y(s) = G(s).X(s) = G(s).1 =

G(s) = k/(τs + 1).

Dividing numerator and denominator

by τ we have:

$(k/\tau)/((\tau s/\tau) + (1/(\tau))) =$

$(k * (1/\tau))/(s + (1/\tau)).$

Using the standard Laplace

transform table we have:

$\frac{k}{\tau}(e^{-(\frac{1}{\tau})t}) =$

$\frac{k}{\tau}(e^{-(\frac{t}{\tau})}).$

The output of a second order

system:

$$\left(\frac{d^2 y}{dt^2}\right) + 2\xi\omega_n\left(\frac{dy}{dt}\right) + \left(\omega_n\right)^2 y$$

$$=$$

$$kx$$

(a non-homogenous equation) where at steady state we have:

$(w_n)^2 y = kx$.

This equation can also be written as:

$$\left(\frac{d^2 y}{dt^2}\right) + 2\xi\omega_n\left(\frac{dy}{dt}\right) + \left(\omega_n\right)^2 y$$

$$=$$

$$k\left(\left(w_n\right)^2\right)x$$

where the steady state is at y = kx.

ξ is the damping ratio and (w_n) is the natural angular frequency at which the system oscillates.

Changing this second order into an

s-function we have:

$$s^2 Y(s) + 2(\xi)(w_n)sY(s) + (w_n)^2 Y(s)$$

$$= k(w_n)^2 X(s)$$

where y (y(t)) is the output and x
(x(t)) is the input to the system.

$Y(s)[s^2 + 2\xi(w_n)s + (w_n)^2] =$
$X(s)k(w_n)^2$

hence

$Y(s)/X(s) =$

$(k(w_n)^2/\{s^2 + 2\xi(w_n)s + (w_n)^2\})$,

$$\frac{Y(s)}{X(s)} = \frac{k(w_n)^2}{s^2 + 2\xi w_n s + (w_n)^2}$$

$G(s) = Y(s)/X(s)$.

Now for a system subjected to a
unit step input function that is
$X(s) = 1/s$ we have
$Y(s) = G(s).X(s) = G(s).(1/s) =$
$[(k(w_n)^2)/\{s^2 + 2\xi(w_n)s +$
$(w_n)^2\}]*(1/s) =$

$$= \frac{k(w_n)^2}{s^2 + 2\xi w_n s + (w_n)^2}\left(\frac{1}{s}\right)$$

$$= k(w_n)^2 / (s(s^2 + 2(\xi)w_n s + (w_n)^2))$$

assuming the roots of

$(s^2 + 2(\xi)w_n s + (w_n)^2) = 0$ are p_1

and p_2 then we have

$k(w_n)^2 / [s(s + p_1)(s + p_2)]$

this can also be written as:

$(p_1 p_2 / p_1 p_2)[k(w_n)^2 / (s(s + p_1)(s +$

$p_2))]$ =

$$\frac{p_1 p_2}{p_1 p_2} \times \frac{k(w_n)^2}{s(s + p_1)(s + p_2)}$$

Using the almighty formula to show

from the equation $s^2 + 2(\xi)(w_n)s +$

$(w_n)^2 = 0$

the actual values of the roots we

have:

$$x = \frac{-b \pm \sqrt{b^2 - 4ac}}{2a}$$

where a = 1

$b = 2(\xi)w_n$

$c = (w_n)^2$

thus

$$p = x =$$

$$\frac{-2\xi\omega_n \pm \sqrt{\left(2\xi\omega_n\right)^2 - 4 \times 1 \times \left(\omega_n\right)^2}}{2 \times 1}$$

$$=$$

$$\frac{-2\xi\omega_n \pm \sqrt{\left(2^2 \xi^2 \left(\omega_n\right)^2\right) - 4 \times \left(\omega_n\right)^2}}{2}$$

$$=$$

$$\frac{-2\xi\omega_n \pm \sqrt{\left(4\xi^2 \left(\omega_n\right)^2\right) - 4 \times \left(\omega_n\right)^2}}{2}$$

$$=$$

$$\frac{-2\xi\omega_n \pm \sqrt{4\left(\xi^2 \left(\omega_n\right)^2 - \left(\omega_n\right)^2\right)}}{2}$$

$$=$$

$$\frac{-2\xi\omega_n \pm \sqrt{4\left(\omega_n\right)^2\left(\xi^2 - 1\right)}}{2}$$

$$=$$

$$\frac{-2\xi\omega_n \pm 2\omega_n\sqrt{\left(\xi^2 - 1\right)}}{2}$$

$$=$$

$$\frac{-2\xi\omega_n}{2} \pm \frac{2\omega_n\sqrt{\left(\xi^2 - 1\right)}}{2}$$

$$=$$

$$-\xi\omega_n \pm \omega_n\sqrt{\left(\xi^2 - 1\right)}$$

thus $p = -\xi\omega_n \pm \omega_n\sqrt{\left(\xi^2 - 1\right)}$

$p_1 = -\xi\omega_n + \omega_n\sqrt{\left(\xi^2 - 1\right)}$

or

$p_2 = -\xi\omega_n - \omega_n\sqrt{\left(\xi^2 - 1\right)}$

Depending on the damping factor ,

ξ, and whether the system is

damped, overdamped or underdamped

we can have three different forms

of answers to the equation,

$Y(s) = G(s)X(s)$

where $X(s) = 1/s$ and hence

$Y(s) =$

$k(w_n)^2/[s(s^2 + 2(\xi)(w_n)s + (w_n)^2)]$.

These different forms are:

1. where damping factor $(\xi) > 1$:

the roots p_1 and p_2 are real and

hence using partial fraction on the

equation

$k(w_n)^2/[s(s + p_1)(s + p_2)]$

and taking the inverse laplace

transform of the result we will

have as the output

y =

$$\frac{k(w_n)^2}{p_1 p_2}\left[1-\left(\frac{p_2}{p_2-p_1}\right)e^{-p_1 t}+\left(\frac{p_1}{p_2-p_1}\right)e^{-p_2 t}\right]$$

the system in this case is

overdamped. Here as t -> ∞ the

exponential approaches to 0 and

thus

y =

$$\frac{k(\omega_n)^2}{p_1 p_2}[1-0+0]$$

$$=$$

$$\frac{k(\omega_n)^2}{p_1 p_2}$$

Now since

$$p_1 \times p_2 = \left(-\xi(\omega_n)+\omega_n\left(\sqrt{\xi^2-1}\right)\right)\left(-\xi(\omega_n)-\omega_n\left(\sqrt{\xi^2-1}\right)\right)$$

$$=$$

$$(\omega_n)^2$$

then y = $(k(w_n)^2)/(w_n)^2$

and thus y = k.

(2) where damping factor (ξ) = 1:

at ξ = 1

the square root term, $\sqrt{((\xi)^2 - 1)}$,

becomes:

$$\sqrt{\xi^2 - 1} =$$

$\sqrt{(1^2 - 1)}$ = $\sqrt{(0)}$ = 0

and hence,

$$p = -\xi\omega_n \pm \omega_n\sqrt{\xi^2 - 1}$$

and thus

$p_1 = p_2 = -1 \times (w_n) = -(w_n)$.

The value of the roots are the

same (that is $(-w_n)$ and thus the

output equation is

$Y(s) = k(w_n)^2/s(s + w_n)^2$

Using partial fraction

$Y(s) = k(w_n)^2/s(s + (w_n))^2 =$

$(A/s) + (B/(s + w_n))$

+ (c/((s + w_n)^2)) =

(A(s + w_n)^2 + B(s + w_n)s + Cs)/(s(s

+ w_n)^2) =

$$\frac{A\left(s^2 + 2s\omega_n + \left(\omega_n\right)^2\right) + Bs\left(s + \omega_n\right) + Cs}{s\left(s + \omega_n\right)^2}.$$

Expanding the numerator and

equating coefficient and constants

we have:

As^2 + 2Asw$_n$ + A(w^2)$_n$ + Bs^2 + Bs(w$_n$)+

Cs

where equating coefficients and

constants we have

k(w$_n$)^2 = A(w$_n$)^2 ..... (1.) ,

k = A

As^2 + Bs^2 = 0 ..... (2.),

A + B = 0 ,

B = -A = -(k) = -k

Cs + 2Asw$_n$ + Bsw$_n$ = 0,

C + 2Aw$_n$ + Bw$_n$ = 0,

C = -2Aw$_n$ - Bw$_n$ = -2kw$_n$ - (-k)w$_n$ =

$-2kw_n + kw_n = kw_n - 2kw_n$

$c = -kw_n$

thus

$(k(w_n)^2)/(s(s + w_n)^2) =$

$(k/s) + (-k)/(s + w_n) +$

$((-kw_n)/((s + w_n)^2)$

Using inverse laplace transform operation we have:

$$y = k\left(1 - e^{-w_n t} - w_n t e^{-w_n t}\right)$$

Now as time (t) tends to ∞ then $e^{-w_n t}$ (the exponential) tends to zero and hence $y = k$. This is the critically damped condition.　　(3)

where damping factor $(\xi) < 1$:

the square root term in this case does not have a real value and thus using standard laplace transform we have:

$$L^{-1}\left\{\frac{k(\omega_n)^2}{s(s^2 + 2\xi\omega_n s + (\omega_n)^2)}\right\}$$

$$= y$$

$$=$$

$$k\left(1 - \left(\frac{1}{\sqrt{1-\xi^2}}\right) \times e^{-\xi\omega_n t} \sin\left(\omega_n\sqrt{1-\xi^2}t\right) + \phi\right)$$

where $\cos(\phi) = \xi$.

This is the underdamped

oscillation. y = k here as t -> ∞

that is as time(t) -> ∞ the

exponential tends to zero (0) and

hence y = k.

Examples:

the input x and output y of a system are described by the differential equation

$$\frac{d^2 y}{dt^2} + 3\frac{dy}{dt} + 2y = x$$

if initially the input and output are zero, what will be the output when there is a unit step input ?

Solution:

(d²y/dt²) + 3(dy/dt) + 2y = x,

L{ (d²y/dt²) + 3(dy/dt) + 2y } =

L{x} = s²Y(s) + 3sY(s) + 2Y(s) =

X(s)

since initial input and output are

zero.

$Y(s)[s^2 + 3s + 2] = X(s)$,

$s^2 + 3s + 2 = X(s)/Y(s)$,

thus $Y(s)/X(s) = 1/(s^2 + 3s + 2)$

since $X(s) = 1/s$ (unit step input)

then

$Y(s) = X(s)/(s^2 + 3s + 2) =$

$X(s)*(1/(s^2 + 3s + 2)) =$

$(1/s)(1/(s^2 + 3s + 2))$

$= 1/(s(s^2 + 3s + 2))$

$= 1/s(s(s + 2) + 1(s + 2))$

$= 1/s(s + 1)(s + 2) =$

$(A/s) + (B/(s + 1)) + (C/(s + 2))$

numerator:

$\{A(s^2 + 3s + 2) + B(s(s + 2)) +$

$C(s(s + 1)) \} =$

$As^2 + 3As + 2A + Bs^2 + 2Bs + Cs^2 +$

Cs

equating coefficients of the

numerator on both sides of the

equation we have:

$As^2 + Bs^2 + Cs^2 = 0$,

$A + B + C = 0$ .... (1.) ,

$3As + 2Bs + Cs = 0$,

$3A + 2B + C = 0$ .... (2.) ,

$2A = 1$ .... (3)

thus $2A = 1$,

$A = 1/2$.

Equation (2) $-$ (1):

$2A + B = 0$,

$B = -2A = -2(1/2) = -1$

also

$C = -A - B = (-1/2) + 1 = 1/2$

hence $1/(s(s + 2)(s + 1)) =$

$((1/2)/s) + (-1)/(s + 1) +$

$(1/2)/(s + 2) =$

$(1/2) + (-e^{-t}) + (1/2)e^{-2t} =$

$1/2 - e^{-t} + (1/2)e^{-2t}$.

a system has a forward path

transfer function of (10/(s + 3))

and a negative feedback loop with

transfer function 5. What is the

time constant of the resulting

first-order system ?

Solution:

G(s) = 10/(s + 3) ,

H(s) = 5

Given:

Y(s)/X(s) = G(s)/(1 + G(s)H(s)) =
(10/(s + 3))/(1 + (10/(s + 3))(5))
= (10/(s + 3))/(1 + (50/(s + 3)))
= (10/(s + 3))/((s + 3 + 50)/(s +
3)) =
(10/(s + 3))*((s + 3)/(s + 53)) =
10/(s + 53) =
10(1/(s + 53))

dividing numerator and denominator by 53 we have:

$= 10((1/53)/((s/53) + (53/53))) = 10((1/53)/((s/53) + 1))$.

Given $G(s) = k/(\tau s + 1)$

comparing:

$\tau s = s/53$

$\tau = (1/53)$ seconds. This is the time constant.

Example3:

a system has a transfer function of $(2/(s + 1))$. What will be its output as a function of time when subjected to (a) a step input of 3V, (b) an impulse input of 3V ?

(a). $Y(s)/X(s) = 2/(s + 1)$,

$Y(s) = (2/(s + 1))*X(s)$

since $X(s) = 3/s$ (that is $L\{3V\} = 3/s$)

thus $Y(s) = (2/(s + 1)).(3/s) =$

$\{2/(s(s + 1))\}.3 =$

$(6/s(s + 1)) = (A/s) + (B/(s + 1))$

$= (A(s + 1) + Bs)/(s(s + 1))$.

Equating coefficients:

$As + Bs = 0$,

$A + B = 0$ ... (1).

$A = 6$

thus $B = -A = -6$

thus $6/(s(s + 1))$

$= (6/s) - (6/(s + 1))$

$= L^{-1}\{(6/s) - (6/(s + 1))\}$

$= 6 - 6e^{-t}$.

(b) $Y(s)/X(s) = 2/(s + 1)$,

$Y(s) = (2/(s + 1))*X(s)$

$X(s) = 3$

hence

$= (2/(s + 1))*3$

$= (6/(s + 1))$

$= 6(1/(s + 1))$

$L^{-1}\{ (6*1)/(s + 1) \} = 6e^{-t}$

Example4:

a system has a transfer function of $(100/(s^2 + s + 100))$. What will be its natural frequency (w_n) and its damping ratio (ξ)?

Solution:

given:

$G(s) =$

$k(w_n)^2 / (s^2 + 2(\xi)(w_n)s + (w_n)^2)$

comparing above with given

transfer function, that is

$100/(s^2 + s + 100)$

$(w_n)^2 = 100$,

$\sqrt{((w_n)^2)} = \sqrt{100} = 10$,

$w_n = 10$

since $2(\pi)f = w_n$

$f = ((w_n)/2(\pi)) = (10/2(\pi)) = 5/(\pi)$

$2(\xi)(w_n)s = s$

$2(\xi)(w_n) = 1$,

$\xi = 1/(2(w_n)) = 1/(2*10) = 0.05$.

Thus given natural frequency (w) =

$w = (w_n)*\sqrt{(1 - ((\xi)^2))}$

then $w = 10*\sqrt{(1 - (0.05)^2)} = 9.987$ approximately 10.

find the transfer function for
each of the following equations
assuming zero initial conditions:

(a) $2(d^2x/dt^2) + dx/dt - x = f(t)$

(b) $3(d^3y/dt^3) + 6(d^2y/dt^2) +$
$8(dy/dt) + 4y = g(t)$

(a) $1/(2s^2 + s - 1)$

(b) $1/(3s^3 + 6s^2 + 8s + 4)$.

CHAPTER SIX

MATHEMATICAL MODELLING

The process by which a physical system is simplified to obtain a mathematically tractable situation is called modelling. The resulting simplified idealised version of the real system whose behaviour approximates that of the real system is called the mathematical model or a model of a system. Fundamental laws, such as Newton's law, law of conversation of mass, kirchhoff voltage laws etc, based on the system involved, are used to obtain mathematical models of systems. These models which for dynamics systems are differential equations are often expressed in the form of a transfer function, a configuration form, an input-

output equation or a state-space
representation.

 We, for the purpose of this
chapter, shall focus on the
mathematical modelling of a
translational mechanical system.
Drawing a freebody diagram is the
first step to obtaining the model
of a mechanical system. This
freebody diagram when correctly
drawn helps one to corectly
analyze the system and obtain a
correct mathematical model of it.
Examples are given below to
illustrate the use of a freebody
diagram:

(1) A simple mechanical system of
mass, spring and damper is shown
below:

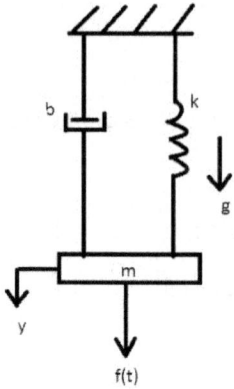

Draw the necessary freebody
diagram and derive the
differential equation.

this is a mechanical system
(translational system) with one
degree of freedom (one co-
ordinate, the displacement y). The
mechanical elements are the mass
(m), spring (k) and damper (b). We
can draw the freebody diagram as
below:
(1) identify (isolate) the forces
at work and their directions with
the given assumption in mind.

F(t), force due to spring k, the
spring force, and force due to
damper, the damper force, are the
forces at work in this case.
(2) apply Newton's law:
we thus, applying the above steps,
have:

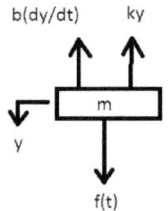

b(dy/dt) = damping force ,
ky = spring force.
Applying Newton's second law:
f(t) - ((b(dy/dt)) + ky) =
m(d^{2}y/dt^2)
hence
f(t) = m(d^{2}y/dt^2) + (b(dy/dt) + ky)
using transfer function to
represent this model we have:
F(s) = ms^{2}Y(s) + bsY(s) + kY(s) ,

$F(s) = Y(s)[ms^2 + bs + k].$

Dividing thro by $Y(s)$:

$F(s)/Y(s) = ms^2 + bs + k,$

$F(s) = input,$

$Y(s) = output$

hence $G(s) = Y(s)/F(s) =$

$1/(ms^2 + bs + k).$

Example2:

---

consider the translational system shown below. Assume that the system is constrained to move in a vertical plane and in a horizontal direction (a) draw the necessary freebody diagrams and derive the differential equations, then (b) put the equations in the second order matrix form:

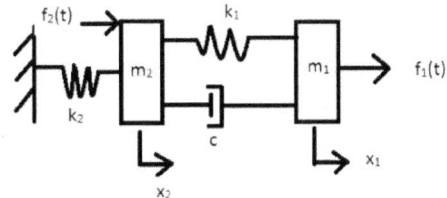

draw the freebody diagram:

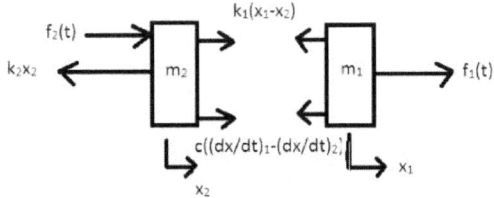

Using Newton's 2nd law:

for m_1 :

collect forces in same direction together:

$f_1(t) - ((k_1(x_1 - x_2)) + (c((dx/dt)_1 - (dx/dt)_2))) = m_1(d^2x/dt^2)_1$

for m_2 :

$f_2(t) + ((k_1(x_1 - x_2)) + c((dx/dt)_1 - (dx/dt)_2)) - k_2x_2 = m_2(d^2x/dt^2)_2$

thus

$f_1(t) = m_1(d^2x/dt^2)_1 + k_1x_1 - k_1x_2 +$

c(dx/dt)$_1$ - c(dx/dt)$_2$

f$_2$(t) = m$_2$(d^{2}x/dt^2)$_2$ + (k$_1$ + k$_2$)x$_2$ -

k$_1$x$_1$ - c(dx/dt)$_1$ + c(dx/dt)$_2$.

The state matrix is:

$$\begin{Bmatrix} m_1, 0 \\ 0, m_2 \end{Bmatrix} \begin{Bmatrix} \ddot{x}_1 \\ \ddot{x}_2 \end{Bmatrix} + \begin{Bmatrix} c, -c \\ -c, c \end{Bmatrix} \begin{Bmatrix} \dot{x}_1 \\ \dot{x}_2 \end{Bmatrix} + \begin{Bmatrix} k_1, -k_1 \\ -k_2, k_1+k_2 \end{Bmatrix} \begin{Bmatrix} x_1 \\ x_2 \end{Bmatrix} = \begin{Bmatrix} f_1 \\ f_2 \end{Bmatrix}$$

Example 3:

consider the two degree of freedom system below:

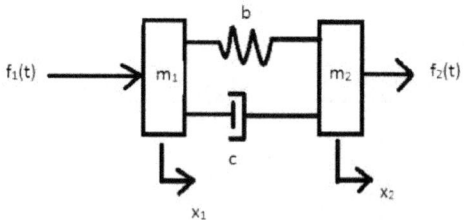

(a) draw the necessary freebody diagram (FBD) and derive the differential equations.

assuming $x_1 > x_2 > 0$ (that is the spring is in compression) and hence $(dx/dt)_1 > (dx/dt)_2 > 0$ the FBD is thus:

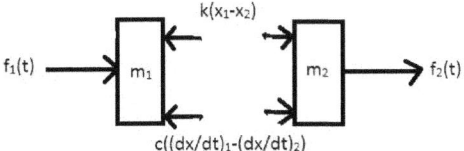

the differential equation is then:
for m_1 :
$f_1(t) - (k(x_1 - x_2) + c((dx/dt)_1 - (dx/dt)_2)) = m_1(d^2x/dt^2)_1$ ....(1.)
for m_2 :
$f_2(t) + k(x_1 - x_2) + c((dx/dt)_1 - (dx/dt)_2) - 0 = m_2(d^2x/dt^2)_2$ ....(2)

making $f_1(t)$ and $f_2(t)$ subject in equ(1) and equ(2) we have:

$f_1(t) = k(x_1 - x_2) + c((dx/dt)_1 - (dx/dt)_2) + m_1(d^2x/dt^2)_1$ ...(3)

$f_2(t) = m_2(d^2x/dt^2)_2 - k(x_1 - x_2) - c((dx/dt)_1 - (dx/dt)_2)$.

Note: note that for a particular or given freebody diagram for a given system $x_1 > x_2 > 0$ or $x_2 > x_1 > 0$, that is a reversion of the damping force and spring force acting between the two masses, $k(x_1 - x_2)$ and $c((dx/dt)_1 - (dx/dt)_2)$ or $k(x_2 - x_1)$ and $c((dx/dt)_2 - (dx/dt)_1)$ respectively will produce same set of differential equations.

consider the translational system shown. Assume that the system is constrained to move in a vertical plane and in horizontal direction. (a) Draw the necessary freebody diagrams and derive the differential equations.

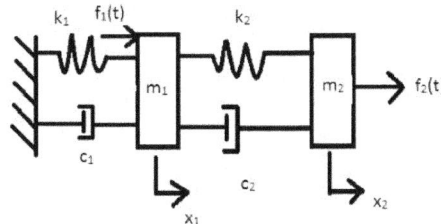

Solution:

$m_1(d^2x/dt^2)_1 = f_1 + c_2((dx/dt)_2$
$- (dx/dt)_1) + k_2(x_2 - x_1)$
$- (c_1(dx/dt)_1 + k_1x_1) \quad .... (1).$
$m_2(d^2x/dt^2) = f_2 - (c_2((dx/dt)_2$
$- (dx/dt)_1) + k_2(x_2 - x_1)).$

APPENDIX OF A MATHEMATICAL
MODELLING PROBLEM AND ANALOG
COMPUTER PROGRAMMING

This appendix is intended to show
clearly, or thoroughly explain,
how the physical laws can be used
to model a dynamic system and how
to simulate this system using
analog computer components.

A MODEL AND A SIMULATION OF A
SIMPLE SPRING, MASS AND DAMPER
SYSTEM

INTRODUCTION

A model of a dynamic system is an
equation or a set of equations
that represents the dynamics of a
system. This equation or equations
are differential equations formed
using the physical laws, such as
Newton's laws, kirchhoff's voltage

and current laws and laws of
conservation of mass, that govern
a particular system. A simulation
of this system, once modelled, can
be done on analog computers.

PROCESSES INVOLVED IN THE
SIMULATION OF A SYSTEM

I will consider the simulation
processes in this section. This
include mathematical model of
system, writing the program,
magnitude and time scaling, static
checking and lastly but will not
be considered, computer
implementation.

*Mathematical modelling of the
system:*

it involves definition of some
physical system or process in
terms of differential equations.
The physical system is here

simplified into a mathematically tractable situation or form called model. This model is a simplified version of the real system. A translational system shown below will be considered in this case for illustration of the modelling principles. A two degree of freedom mechanical system subjected to two applied forces:

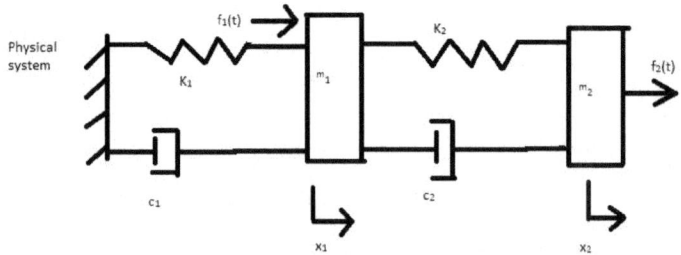

free body diagram :

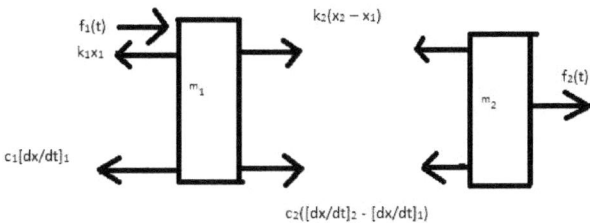

Where spring force (f) = kx, damping/viscous friction force = c(dx/dt). Where c = damper (a device that provides viscous friction) coefficient.

K = spring constant.

Deriving equation for each mass:

```
f₁(t)  +  k₂( x₂ -  x₁ )  +  c₂(  [dx/dt]₂
-  [dx/dt]₁ )  -  (k₁x₁ +  c₁[dx/dt]₁)  =
m₁[d²x/(dt)²]₁ ,
f₂(t)     -    (    k₂    (x₂    -    x₁)    +
c₂ (   [dx/dt]₂  -   [dx/dt]₁ )  )     =
m₂[d²x/(dt)²]₂.
```

Dot representation of the above equation:

$$f_1(t) + k_2(x_2 - x_1) + c_2\left(\dot{x}_2\right) - \left(\dot{x}_1\right) - (k_1 x_1 + c_1\left(\dot{x}_1\right))$$

$$= m_1\left(\ddot{x}_1\right)$$

$$f_2(t) - (k_2(x_2 - x_1) + c_2\left(\dot{x}_2\right) - \left(\dot{x}_1\right))$$

$$= m_2\left(\ddot{x}_2\right)$$

Writing the program:
this involves the manipulation of

the mathematical equation (model) using some basic analog components. These components which can be linear or non-linear help to relate the sequence of mathematical operations in the equation. The linear components perform mathematical operations like:

1. *Multiplication by a constant:*

this operation is performed by an analog component called the potentiometer. An example of the use of this component to manipulate this operation is shown below:

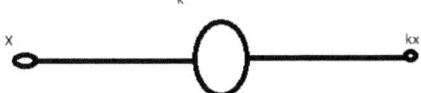

2. *Inversion:*

The analog component involved here is the inverter. This inverts the sign of an input to the system. An example of this operation is as shown below:

Diagram of Inverter:

3. *Algebraic summation:*

the summer is the analog component used in this case. This performs two operations at the same time. It inverts and adds algebraically. An example of this operation is as shown below:

given,

d(dx/dt)/dt + 0.5(dx/dt) + x = 3

that is $\dfrac{d^2x}{dt^2} + 0.5\dfrac{dx}{dt} + x = 3$

Applying summer to represent the equation,

d²x/dt² = - 0.5 dx/dt - x + 3,

= - (0.5$\dot{x}$ + x - 3),

= - (0.5$\dot{x}$ + x - 3) ,   we thus have :

The summer diagram representing the equation :

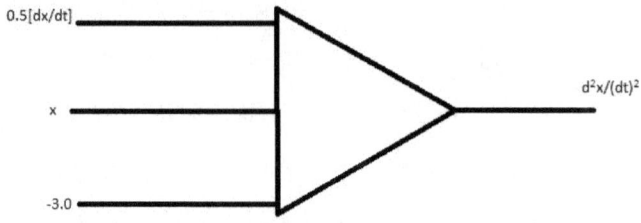

0.5[dx/dt]

x

-3.0

$d^2x/(dt)^2$

(This component inverts each of the input to it to give an exact representation of the output (d(dx/dt)/dt) as it is indicated in the model).

4. *Continuous integration:*

the integrator is the analog component involved in this case. It performs three operations, integration, summation and inversion. An example of the use of this component is shown below using the equation above,

$$d^2x/dt^2 + 0.5\ dx/dt + x = 3$$

Dot representation of the above equation:

$$\ddot{x} + 0.5\,\dot{x} + x = 3$$

Represention of its operation:
$$d^2x/dt^2 = -0.5\ dx/dt - x + 3$$
$$= d(dx/dt)/dt = -0.5dx/dt - x + 3,$$
<the full integrator diagram for the equation>:

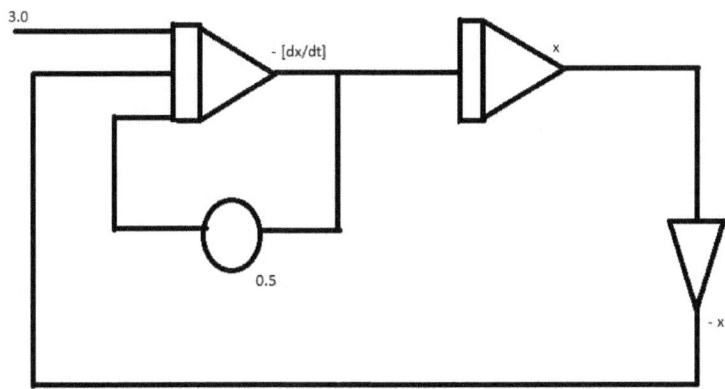

The above operations are enough to solve linear differential equations with constant coefficients of the general form: $a_n(d^nx/dt^n)^m + a_{n-1}(d^{n-1}x/dt^{n-1}) + \ldots a_ox = f(t)$, where all of the a_n's are constant and all of the derivatives of different orders have a degree of $m = 1$.

Equations where $m \neq 1$ and a_n coefficients are not constants are considered to be non-linear differential equations. Non-linear components are used to solve these non-linear differential equations. These components perform the following operations:

1. Multiplication and division of variables,

2. The generation of arbitrary functions, and

3. The mechanization of constraints and elementary logic operations. The non-linear components together with the linear components are used to simulate the non-linear systems that occur in practise.

A simulation of the modelled translational system will be done in this section to demonstrate this stage in simulation. The simulation is as below:

$f_1(t) + k_2(x_2 - x_1) + c_2([dx/dt]_2 - [dx/dt]_1) - (k_1 x_1 + c_1 [dx/dt]_1) = m_1 [d^2 x/(dt)^2]_1 \ldots (1.)$

Dot notation of the above equation:

$$f_1(t) + k_2(x_2 - x_1) + c_2(\dot{x}_2 - \dot{x}_1) - (k_1 x_1 + c_1 \dot{x}_1) = m_1 \ddot{x}_1 \ldots (1.)$$

f$_2$(t)   -   (k$_2$ (x$_2$ -   x$_1$)   +
c$_2$ ([dx/dt]$_2$ - [dx/dt]$_1$))   =
m$_2$[d^{2}x/(dt)^2]$_2$ .... (2.)
Dot   notation   of   the   above
equation:

$$f_2(t) - (k_2(x_2 - x_1) + c_2(\dot{x_2} - \dot{x_1})) = m_2\,\ddot{x_2} \,....(2.)$$

Dividing thro by m$_1$:
(m$_1$([d^{2}x/dt^2]$_1$))/m$_1$ =

f$_1$(t)/m$_1$ +   (k$_2$/m$_1$)(x$_2$ -   x$_1$)   +
(c$_2$/m$_1$)([dx/dt]$_2$ -   [dx/dt]$_1$)
- ((k$_1$x$_1$ + c$_1$[dx/dt]$_1$)/m$_1$)
=   (f$_1$(t)/m$_1$)+(k$_2$/m$_1$)(x$_2$ -   x$_1$)   +
(c$_2$/m$_1$)([dx/dt]$_2$ -   [dx/dt]$_1$)   -
(k$_1$/m$_1$)x$_1$ - (c$_1$/m$_1$)[dx/dt]$_1$,
making [d^{2}x/dt^2]$_1$ subject.

A dot notation of the above equation:

$$\frac{m_1 \ddot{x}_1}{m_1} = \frac{f_1(t)}{m_1} + \frac{k_2(x_2 - x_1)}{m_1} + \frac{c_2(\dot{x}_2 - \dot{x}_1)}{m_1} - \frac{(k_1 x_1 + c_1 \dot{x}_1)}{m_1}$$

$$=$$

$$\frac{f_1(t)}{m_1} + \frac{k_2(x_2 - x_1)}{m_1} + \frac{c_2(\dot{x}_2 - \dot{x}_1)}{m_1} - \frac{k_1 x_1}{m_1} - \frac{c_1 \dot{x}_1}{m_1}$$

Also,

$m_2 [d^2x/(dt)^2]_2. = f_2(t) - k_2 (x_2 - x_1) - c_2 ([dx/dt]_2 - [dx/dt]_1)$,

Or

$$m_2 \ddot{x}_2 = f_2(t) - (k_2(x_2 - x_1) + c_2(\dot{x}_2 - \dot{x}_1))$$

$$= f_2(t) - k_2(x_2 - x_1) - c_2(\dot{x}_2 - \dot{x}_1)$$

dividing thro by m₂ =

$(m_2/m_2) [d^2x/(dt)^2]_2$ =

$(f_2(t)/m_2)$ — $(k_2/m_2)(x_2 - x_1)$ —

$(c_2/m_2)([dx/dt]_2 - [dx/dt]_1)$,

Or,

$$\frac{m_2 \ddot{x}_2}{m_2} = \frac{f_2(t)}{m_2} - \frac{k_2(x_2 - x_1)}{m_2} - \frac{c_2(\dot{x}_2 - \dot{x}_1)}{m_2}$$

making $[d^2x/(dt)^2]_2$ the subject of the equation .

Next, form the integrator strings for the equations with $[d^2x/dt^2]_1$ and $[d^2x/(dt)^2]_2$ as subjects. The highest order derivative for each of these equations is two. Thus I generate an integrator string for each of the equations having

-dx/dt and x as outputs. A positive x, note, is needed as the final output. The integrator string is thus as below:

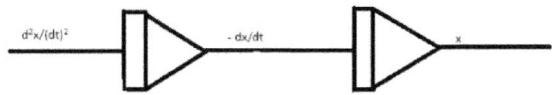

This implies for the 1st integrator:

$-\int [d^2x/(dt)^2] = - \, dx/dt$

and for the last integrator:

$-\int -dx/dt = \int dx/dt = x$,

a simulation of the system with initial conditions $[dx/dt]_1(0)$,.

$[dx/dt]_2(0)$, $x_1(0)$ and $x_2(0)$ using all necessary components:

<the diagram: see page 239 >

Magnitude and Time scaling techniques:

Magnitude and time scaling arise because of the incompatibility of analog computer and natural phenomena. The analog computer can only vary between plus one and minus one machine unit ($-1 \leq x \leq 1$) but natural phenomena can be very large or exceedingly small such that man may not be able to interpret it. Thus a problem must be scaled so that it will fit the

computer, and there will be no overload where a problem variable existing as an amplifier (e.g. Integrator) output is so large to cause the amplifier, the basic

unit of analog computer that is
integrators, summers, inverters
etcetra that can be used, to
overload. The program of the
translational system above will be
used below to illustrate these
techniques:

*the amplifier outputs of the
system in the program include:*
$[dx/dt]_1$, x_1 , $[dx/dt]_2$ and x_2 ,
$x_2 - x_1$, $[dx/dt]_2 - [dx/dt]_1$ (also
called problem variables).
Thus, to make the variation of
these outputs to fall within one
machine unit, each of them will be
divided by their maximum value.
That is,

$| x_1/x_{1max} | \leq 1,$

$| [dx/dt]_1 / [dx/dt]_{1max} | \leq 1,$
$| x_2/x_{2max} | \leq 1,$
$| [dx/dt]_2 / [dx/dt]_{2max} | \leq 1,$
$| (x_2 - x_1) / (x_2 - x_1)_{max} | \leq 1$
and

$|[dx/dt]_2-[dx/dt]_1/([dx/dt]_2-[dx/dt]_1)_{max}| \le 1.$

$$\left|\frac{x_1}{(x_1)_{max}}\right| \le 1,$$

$$\left|\frac{\dot{x}_1}{(\dot{x}_1)_{max}}\right| \le 1,$$

$$\left|\frac{x_2}{(x_2)_{max}}\right| \le 1,$$

$$\left|\frac{\dot{x}_2}{(\dot{x}_2)_{max}}\right| \le 1,$$

$$\left|\frac{x_2 - x_1}{(x_2 - x_1)_{max}}\right| \le 1,$$

$$\left|\frac{(\dot{x}_2)-(\dot{x}_1)}{((\dot{x}_2)-(\dot{x}_1))_{max}}\right| \le 1$$

Hence if these are made outputs of the amplifiers instead of x_1, $[dx/dt]_1$, x_2, $[dx/dt]_2$, x_2-x_1 and $[dx/dt]_2 - [dx/dt]_1$ the problem of magnitude will be solved. Note that in magnitude scaling the outputs of all amplifiers must be kept within minus one and plus one machine unit. Thus from equations (1.) and (2.) above the outputs are $[dx/dt]_1$, x_1, $[dx/dt]_2$, x_2, x_2-x_1 and $[dx/dt]_2 - [dx/dt]_1$ and hence to account for their maximum values, and keep the amplifier outputs within plus one and minus one machine units we do the following,

for equation 1: $[d^2x/dt^2]_1 =$
$(f_1(t)/m_1) + (k_2/m_1)(x_2 - x_1) +$
$(c_2/m_1)([dx/dt]_2 - [dx/dt]_1)$
$- ((k_1/m_1)x_1 + (c_1/m_1)[dx/dt]_1),$

$$\ddot{x}_1 = \frac{f_1(t)}{m_1} + \frac{k_2(x_2 - x_1)}{m_1} + \frac{c_2(\dot{x}_2 - \dot{x}_1)}{m_1} - (\frac{k_1 x_1}{m_1} + \frac{c_1 \dot{x}_1}{m_1})$$

applying the values to keep outputs within plus one and minus one machine units and at the same time maintain the equation as above we have:

(d(|[dx/dt]$_1$/[dx/dt]$_{1max}$|)/dt)*([dx/dt]$_{1max}$) = d([dx/dt]$_1$)/dt

= [d^{2}x/dt^2]$_1$,

(k$_2$/m$_1$)(|(x$_2$-x$_1$)/(x$_2$-x$_1$)$_{max}$|)*(x$_2$-x$_1$)$_{max}$ =

(k$_2$/m$_1$)(x$_2$-x$_1$),

(c$_2$/m$_1$)*(|[dx/dt]$_2$-[dx/dt]$_1$/([dx/dt]$_2$-[dx/dt]$_1$)$_{max}$|)*([dx/dt]$_2$-[dx/dt]$_1$)$_{max}$ =

$(c_2/m_1)*([dx/dt]_2-[dx/dt]_1)$,

$(k_1/m_1)(|x_1/x_{1max}|)*x_{1max} = (k_1/m_1)*x_1$,

$(c_1/m_1)*(|[dx/dt]_1/[dx/dt]_{1max}|)$

$*[dx/dt]_{1max}| = (c_1/m_1)*[dx/dt]_1$,

Or,

$$\frac{d\left(\frac{\dot{x}_1}{\left(\dot{x}_1\right)_{max}}\right)\bullet\left(\left(\dot{x}_1\right)_{max}\right)}{dt} = \frac{d\left(\dot{x}_1\right)}{dt} = \ddot{x}_1,$$

$$\frac{k_2}{m_1}\left(\frac{x_2-x_1}{(x_2-x_1)_{max}}\right)\bullet(x_2-x_1)_{max} = \frac{k_2}{m_1}(x_2-x_1),$$

$$\frac{c_2}{m_1}\frac{\left(\dot{x}_2-\dot{x}_1\right)}{\left(\dot{x}_2-\dot{x}_1\right)_{max}}\bullet\left(\dot{x}_2-\dot{x}_1\right)_{max} = \frac{c_2\left(\dot{x}_2-\dot{x}_1\right)}{m_1},$$

$$\frac{k_1}{m_1}\frac{x_1}{(x_1)_{max}}\bullet(x_1)_{max} = \frac{k_1 x_1}{m_1},$$

$$\frac{c_1}{m_1}\frac{\dot{x}_1}{\left(\dot{x}_1\right)_{max}}\left(\dot{x}_1\right)_{max} = \frac{c_1\dot{x}_1}{m_1}$$

thus equation (1) becomes:

```
(d(|[dx/dt]₁
/[dx/dt]₁ₘₐₓ|)/dt)*([dx/dt]₁ₘₐₓ) =
(f₁(t)/m₁) +
(k₂/m₁)|(x₂-x₁)/(x₂-x₁)ₘₐₓ|*(x₂-x₁)ₘₐₓ
```

```
+       (c₂/m₁)*|[dx/dt]₂-[dx/dt]₁/
([dx/dt]₂-[dx/dt]₁)max|  *  ([dx/dt]₂-
[dx/dt]₁)max -
(k₁/m₁) | x₁/x₁max|  * x₁max - (c₁/m₁)*(
|[dx/dt]₁ /[dx/dt]₁max|)*[dx/dt]₁max ,
```

Or,

$$\frac{d\left|\dfrac{\dot{x}_1}{\left(\dot{x}_1\right)_{max}}\right|\left(\dot{x}_1\right)_{max}}{dt} = \frac{f_1(t)}{m_1} + \frac{k_2}{m_1}\left|\frac{(x_2-x_1)}{(x_2-x_1)_{max}}\right|\bullet(x_2-x_1)_{max} + \frac{c_2}{m_1}\left|\frac{\dot{x}_2-\dot{x}_1}{\left(\dot{x}_2-\dot{x}_1\right)_{max}}\right|\bullet\left(\dot{x}_2-\dot{x}_1\right)_{max} - \frac{k_1}{m_1}\left|\frac{x_1}{x_{1max}}\right|\bullet x_{1max} - \frac{c_1}{m_1}\left|\frac{\dot{x}_1}{\left(\dot{x}_1\right)_{max}}\right|\left(\dot{x}_1\right)_{max}$$

dividing all through by [dx/dt]₁max
we have :

```
(d(|[dx/dt]₁/[dx/dt]₁max|)/dt)*([dx/
dt]₁max)*(1/[dx/dt]₁max)
=       (1/[dx/dt]₁max)*(f₁(t)/m₁) +
```

$(1/[dx/dt]_{1max}) * (k_2/m_1) | (x_2-x_1)/(x_2-x_1)_{max}| * (x_2-x_1)_{max} +$

$(1/[dx/dt]_{1max}) * (c_2/m_1) * | [dx/dt]_2 - [dx/dt]_1/([dx/dt]_2-[dx/dt]_1)_{max}| * ([dx/dt]_2-[dx/dt]_1)_{max} -$

$(1/[dx/dt]_{1max}) * (k_1/m_1) |x_1/x_{1max}| * x_{1max} - (1/[dx/dt]_{1max}) * (c_1/m_1) * (| [dx/dt]_1/[dx/dt]_{1max}|) * [dx/dt]_{1max}$,

Or,

$$\left(\frac{d\left|\frac{x_1}{\left(\dot{x}_1\right)_{max}}\right|}{dt}\right)\frac{\left(\dot{x}_1\right)max}{\left(\dot{x}_1\right)max} = \left(\frac{1}{\left(\dot{x}_1\right)max}\right)\left|\frac{f_1(t)}{m_1}+\frac{k_2}{m_1}\left|\frac{x_2-x_1}{(x_2-x_1)max}\right|\bullet\frac{(x_2-x_1)max}{\left(\dot{x}_1\right)max}+\frac{c_2}{m_1}\left|\frac{\dot{x}_2-\dot{x}_1}{\left(\dot{x}_2-\dot{x}_1\right)max}\right|\frac{\left(\dot{x}_2-\dot{x}_1\right)max}{\left(\dot{x}_1\right)max}\right.$$

$$\left.-\frac{k_1}{m_1}\left|\frac{x_1}{x_1max}\right|\bullet\frac{x_1max}{\left(\dot{x}_1\right)max}-\frac{c_1}{m_1}\left|\frac{\dot{x}_1}{\dot{x}_1max}\right|\bullet\frac{\dot{x}_1max}{\dot{x}_1max}\right.$$

the amplifier outputs are now:

$| (x_2-x_1)/ (x_2-x_1)_{max}|$,

$|[dx/dt]_2-[dx/dt]_1/([dx/dt]_2-[dx/dt]_1)_{max}|$,

$|x_1/x_{1max}|$

and $|[dx/dt]_1/[dx/dt]_{1max}|$

Or,

$$\frac{x_2 - x_1}{(x_2 - x_1)\max},$$

$$\frac{\dot{x}_2 - \dot{x}_1}{\left(\dot{x}_2 - \dot{x}_1\right)\max},$$

$$\frac{x_1}{x_1\max},$$

$$\frac{\dot{x}_1}{\dot{x}_1\max}$$

(these are the computer variables)
and their coefficients are :

$(x_2-x_1)_{max}*(1/[dx/dt]_{1max})$,

$([dx/dt]_2-[dx/dt]_1)_{max}*$

212

$(1/[dx/dt]_{1max})$

and $x_{1max} * (1/[dx/dt]_{1max})$.

Or,

$$\frac{(x_2 - x_1)\,\text{max}}{\left(\dot{x}_1\right)\text{max}},$$

$$\frac{(\dot{x}_2 - \dot{x}_1)\,\text{max}}{\left(\dot{x}_1\right)\text{max}},$$

$$\frac{x_1\,\text{max}}{\left(\dot{x}_1\right)\text{max}}$$

The product of each of these coefficients represents for each amplifier the amplifier gain and the pot (potentiometer) setting.

The amplifier output determines the magnitude scaling. The coefficient determines the time scaling. Time scaling is done when the value of any of these coefficients fail to fall within the interval 0.05 (approximately 0.1) and 10 ($0.05 \leq$ coefficient of amplifier output $\leq$ 10). It is important to either speed up or slow down the computer. Thus it, time scaling, allows things to happen in a reasonable amount of time such that the human operator can completely observe the phenomena.

A technique used for time scaling is to define a constant B such

that it relates the problem time and the computer time. That is

$$\beta = \frac{computer\,(machine)\,time}{\Pr oblem\,time} = \frac{\tau}{t}$$

The unit of β is the unit of computer time to that of the problem time. Its magnitude indicates the factor by which the problem is sped up or slowed down. Where $\beta > 1$, computer time is large or slower than problem time but where $\beta < 1$, computer time is smaller or faster than problem time.

The integrator is the only analog component affected by time-scaling, other components are not. This device to accomplish time scaling will need to integrate (accept imput for integration   )

with respect to computer time, not problem time, to produce output in computer time. An expression, to help show this, is derived below :

$\beta = \tau / t$,

dx/dt = dx/dt ,

where $\beta = \tau / t$,

$(\beta * t) / \beta = \tau / t * t / \beta = \tau / \beta$,

thus $dx / d (\tau / \beta) = dx/dt$,

$dx / (d (\tau) * (1 / \beta)) = dx/dt$,

$\beta * dx / d (\tau) = dx/dt$,

$dx / d (\tau) = dx/dt * (1 / \beta)$,

hence inputing dx/dt * $(1 / \beta)$ into an integrator will result in having x(computer time), a variable which is a function of the computer time (its symbol, τ), as output. That is,

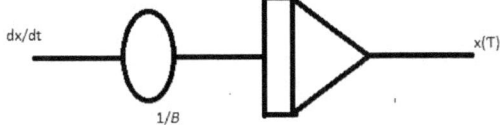

Since ,   x(t) = ∫ (dx/dt

= dx/dt = (dx/dt)(t)) = x(t) ,

hence x(τ) =

∫(((dx/dt)*(1/β)) =   dx/d(τ))

= (dx/d(τ))*(τ) = x(τ) .

Therefore the general rule is that for time scaling all integrator input must be multiplied by 1/β. The translational system will further help to illustrate, as below, this time scaling:

for equation 1 (already magnitude scaled) :

d(| [dx/dt]$_1$ /[dx/dt]$_{1max}$|)/dt =

(1/[dx/dt]$_{1max}$)*(f$_1$(t)/m$_1$)+

(1/[dx/dt]$_{1max}$)*(k$_2$/m$_1$)|(x$_2$-x$_1$)/(x$_2$-x$_1$)$_{max}$|*(x$_2$-x$_1$)$_{max}$ +

(1/[dx/dt]$_{1max}$)*(c$_2$/m$_1$)*|[dx/dt]$_2$ -

$[dx/dt]_1/([dx/dt]_2-[dx/dt]_1)_{max}| *$

$([dx/dt]_2-[dx/dt]_1)_{max}$ $-$

$(1/[dx/dt]_{1max}) * (k_1/m_1) |x_1/x_{1max}| *$

x_{1max} $-(c_1/m_1) * (|[dx/dt]_1$

$/[dx/dt]_{1max}|)$,

Or,

$$\frac{d\left(\left|\dfrac{\dot{x}_1}{\left(\dot{x}_1\right)_{max}}\right|\right)}{dt} =$$

$$\left(\frac{1}{\left(\dot{x}_1\right)_{max}}\right)\frac{f_1(t)}{m_1} + \frac{k_2}{m_1}\left|\frac{x_2-x_1}{(x_2-x_1)_{max}}\right| \bullet \frac{(x_2-x_1)_{max}}{\left(\dot{x}_1\right)_{max}} + \frac{c_2}{m_1}\left|\frac{\dot{x}_2-\dot{x}_1}{\left(\dot{x}_2-\dot{x}_1\right)_{max}}\right|\frac{\left(\dot{x}_2-\dot{x}_1\right)_{max}}{\left(\dot{x}_1\right)_{max}}$$

$$-\frac{k_1}{m_1}\left|\frac{x_1}{x_1\,max}\right| \bullet \frac{x_1\,max}{\left(\dot{x}_1\right)_{max}} - \frac{c_1}{m_1}\left|\frac{\dot{x}_1}{\dot{x}_1\,max}\right|$$

multiplying both sides by (1/*B*)
will give :

$$(1/B) * (d(|[dx/dt]_1$$
$$/[dx/dt]_{1max}|)/dt) =$$
$$(d(|[dx/dt]_1/[dx/dt]_{1max}|)/dT) =$$
$$(1/B) * (1/[dx/dt]_{1max}) * (f_1(t)/m_1) +$$
$$(1/B) * (1/[dx/dt]_{1max}) * (k_2/m_1) | (x_2 -$$
$$x_1)/(x_2 - x_1)_{max} | * (x_2 - x_1)_{max} +$$
$$(1/B) * (1/[dx/dt]_{1max}) * (c_2/m_1) *$$
$$|[dx/dt]_2 - [dx/dt]_1/([dx/dt]_2 -$$
$$[dx/dt]_1)_{max} | * ([dx/dt]_2 - [dx/dt]_1)_{max}$$
$$- (1/B) * (1/[dx/dt]_{1max}) * (k_1/m_1) |$$
$$x_1/x_{1max} | * x_{1max} - (1/B) * (c_1/m_1) * (|$$
$$[dx/dt]_1 / [dx/dt]_{1max}|) \quad ,$$

Or,

$$\frac{1}{\beta}\left(\frac{d\left|\dfrac{\dot{x}_1}{(\dot{x}_1)_{max}}\right|}{dt}\right) =$$

$$\left(\frac{d\left|\dfrac{\dot{x}_1}{(\dot{x}_1)_{max}}\right|}{d\tau}\right) =$$

$$\frac{1}{\beta}\left(\frac{1}{(\dot{x}_1)_{max}}\right)\left|\frac{f_1(t)}{m_1} + \frac{k_2}{\beta m_1}\left|\frac{x_2-x_1}{(x_2-x_1)_{max}}\right| \bullet \frac{(x_2-x_1)_{max}}{(\dot{x}_1)_{max}} + \frac{c_2}{\beta m_1}\left|\frac{\dot{x}_2-\dot{x}_1}{(\dot{x}_2-\dot{x}_1)_{max}}\right| \frac{(\dot{x}_2-\dot{x}_1)_{max}}{(\dot{x}_1)_{max}}\right.$$

$$\left. -\frac{k_1}{\beta m_1}\left|\frac{x_1}{x_1\,max}\right| \bullet \frac{x_1\,max}{(\dot{x}_1)_{max}} - \frac{c_1}{\beta m_1}\left|\frac{\dot{x}_1}{\dot{x}_1\,max}\right|\right.$$

for equation 2:

[d²x/(dt)²]₂ =

(f₂(t)/m₂)   -   (k₂/m₂)(x₂ - x₁) -
(c₂/m₂)([dx/dt]₂ - [dx/dt]₁),
(d(|[dx/dt]₂

/[dx/dt]$_{2max}$|)/dt)*[dx/dt]$_{2max}$| =

(f$_2$(t)/m$_2$) - (k$_2$/m$_2$)*| (x$_2$-x$_1$)/ (x$_2$-

x$_1$)$_{max}$|*(x$_2$-x$_1$)$_{max}$ -

(c$_2$/m$_2$)(|[dx/dt]$_2$-[dx/dt]$_1$/

([dx/dt]$_2$-[dx/dt]$_1$)$_{max}$|)*([dx/dt]$_2$-

[dx/dt]$_1$)$_{max}$|

Or,

$$\ddot{x}_2 = \frac{f_2(t)}{m_2} - \frac{k_2(x_2 - x_1)}{m_2} - \frac{c_2(\dot{x}_2 - \dot{x}_1)}{m_2},$$

$$\frac{d\left|\frac{\dot{x}_2}{\dot{x}_2\max}\right|\dot{x}_2\max}{dt} =$$

$$\frac{f_2(t)}{m_2} - \frac{k_2}{m_2}\left|\frac{(x_2 - x_1)}{(x_2 - x_1)\max}\right|(x_2 - x_1)\max - \frac{c_2}{m_2}\left|\frac{(\dot{x}_2 - \dot{x}_1)}{(\dot{x}_2 - \dot{x}_1)\max}\right|(\dot{x}_2 - \dot{x}_1)\max$$

dividing thro by [dx/dt]$_{2max}$:

(d(|[dx/dt]$_2$/[dx/dt]$_{2max}$|)/dt)   =

(1/[dx/dt]$_{2max}$)*(f$_2$(t)/m$_2$)   -

$(1/[dx/dt]_{2max}) * (k_2/m_2) * |(x_2-x_1)/$
$(x_2-x_1)_{max}| * (x_2-x_1)_{max} - (1/[dx/dt]_{2max}$
$) * (c_2/m_2) (|[dx/dt]_2-[dx/dt]_1/$
$([dx/dt]_2-[dx/dt]_1)_{max}|) * ([dx/dt]_2-$
$[dx/dt]_1)_{max}.$

Or,

$$\frac{d \left| \dfrac{\dot{x}_2}{\dot{x}_2\,\text{max}} \right|}{dt} =$$

$$\frac{1}{\dot{x}_2\,\text{max}} \frac{f_2(t)}{m_2} - \frac{k_2}{m_2} \left| \frac{(x_2-x_1)}{(x_2-x_1)\text{max}} \right| \frac{(x_2-x_1)\text{max}}{\dot{x}_2\,\text{max}} - \frac{c_2}{m_2} \left| \frac{(\dot{x}_2-\dot{x}_1)}{(\dot{x}_2-\dot{x}_1)\text{max}} \right| \frac{(\dot{x}_2-\dot{x}_1)\text{max}}{\dot{x}_2\,\text{max}}$$

Multiplying all through (both sides) by 1/B for time scaling :

$$(d(|[dx/dt]_2 /[dx/dt]_{2max}|)/dt)*(1/B) = (1/B)*(1/[dx/dt]_{2max})*(f_2(t)/m_2) - (1/B)*(1/[dx/dt]_{2max})*(k_2/m_2)*|(x_2-x_1)/(x_2-x_1)_{max}|*(x_2-x_1)_{max} - (1/B)*(1/[dx/dt]_{2max})*(c_2/m_2)(|[dx/dt]_2-[dx/dt]_1/([dx/dt]_2-[dx/dt]_1)_{max}|)*([dx/dt]_2-[dx/dt]_1)_{max}.$$

Or,

$$\frac{1}{\beta}\frac{d\left|\dfrac{\dot{x}_2}{\dot{x}_2\,\max}\right|}{dt} =$$

$$\frac{1}{\dot{x}_2\,\max}\frac{f_2(t)}{m_2\beta} - \frac{k_2}{m_2\beta}\left|\frac{(x_2-x_1)}{(x_2-x_1)\max}\right|\left|\frac{(x_2-x_1)\max}{\dot{x}_2\,\max}\right| - \frac{c_2}{m_2\beta}\left|\frac{(\dot{x}_2-\dot{x}_1)}{(\dot{x}_2-\dot{x}_1)\max}\right|\left|\frac{(\dot{x}_2-\dot{x}_1)\max}{\dot{x}_2\,\max}\right|$$

The full scaled (magnitude and time scaled) program then becomes:
<The full scaled diagram>

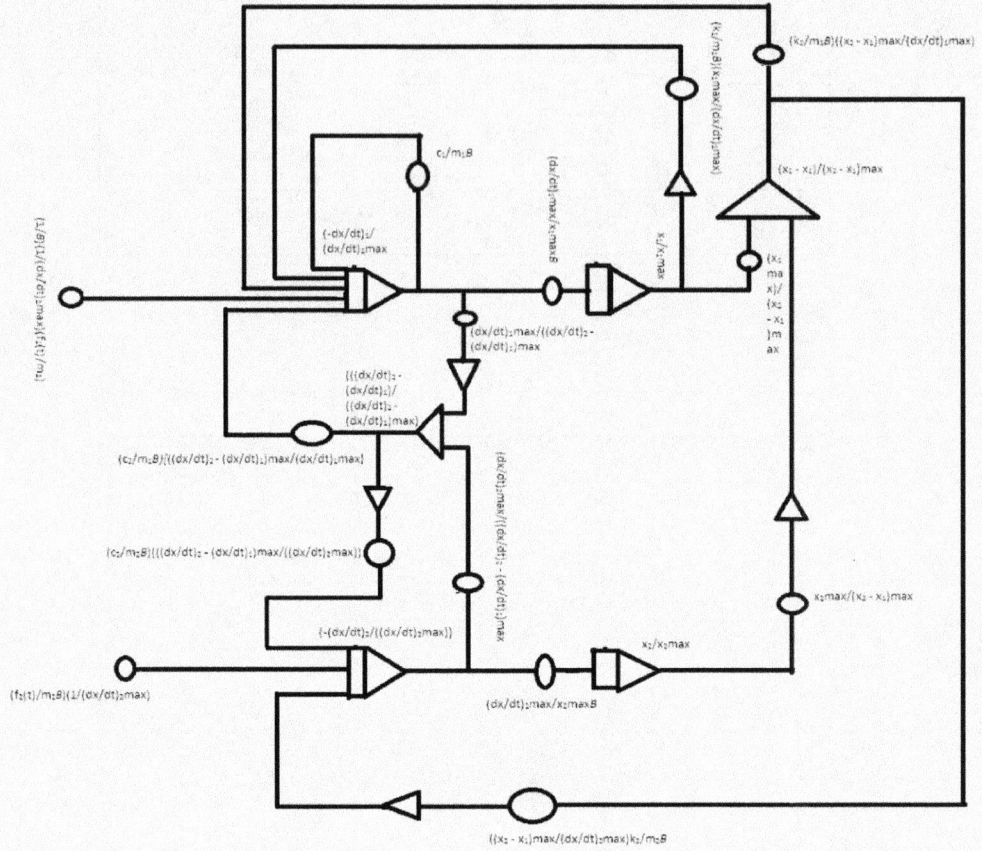

See also pages 240-244 for the full scaled diagram expressed in dot notation.

Static checking:

this is the method of detecting and correcting a program. This involves two parts:
1. a program check,
2. a circuit check.

A program check:

this involves calculations to help detect and correct errors. A separate calculation based on the program and another calculation based on the original problem are performed. The results of each of these calculations are then compared to know if they will be correct, that is give the same results. Calculations based on the program are possible where outputs (computer variable outputs) of all integrator and pot settings are known. The values of the integrator outputs are calculated using the values for the problem variables:

x_1 ,

x_2 ,

$[dx/dt]_1$,

$[dx/dt]_2$

in the problem statement or the modelled equation. The values of the pot setting, whereas, are calculated from the values for the parameters (constant coefficient of these problem variables in the modelled equation). Thus with these values all integrator outputs as well as all derivative inputs to the integrators can be found.

Calculations for the amplifier outputs and their derivatives based on the original problem are also possible. This one can do where one knows the amplifier outputs (the computer variable outputs) and the values given for the problem variables in the original modelled equation of the problem statement.

A circuit check:

this involves patching the program on the computer or connecting from the analog block diagram the electrical components together and measuring voltages to compare it to the calculations.

A SIMULATION OF A SIMPLE SPRING, MASS AND DAMPER SYSTEM

The simulation of this system going through each of the steps involved is as below:

1. *Mathematical modelling:*

a spring, mass and damper system
with a force applied:

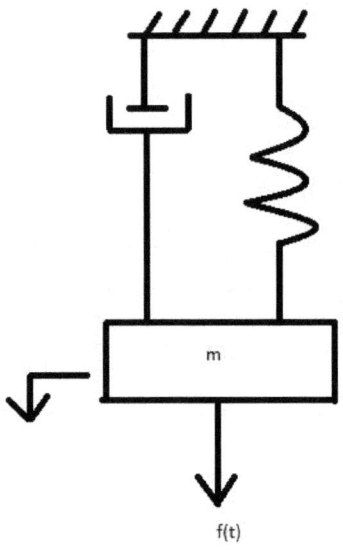

f(t)

A free body diagram of this system to isolate/show each force acting on the single mass (m):

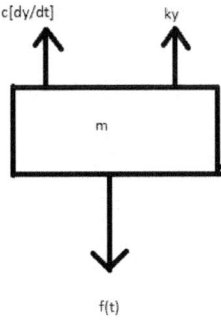

f(t) - (ky + c[dy/dt]) = m[d²y/(dt)²] .... (1.)

Making $d^2y/(dt)^2$ subject of the equation we have:

$d^2y/(dt)^2$ = $(f(t)/m)$ - $((ky + c[dy/dt])/m)$ = $(f(t)/m)$ - ky/m - $c[dy/dt]/m$.

Note using laplace transform for the purpose of relating the output to the input in transfer function this can also be written as :

$F(s)$ - $[kY(s) + csY(s)]$ = $ms^2Y(s)$

or in a state equation and state-space matrix equation form where state variables are x_1 and x_2 and

$x_1 = y$,

$x_2 = dy/dt$,

$[dx/dt]_1 = dy/dt = x_2$,

differentiating x_2 with respect to t :

we have $[dx/dt]_2 = d^2y/dt^2$

and thus the state equations:

$$[dx/dt]_1 = x_2 \quad,$$

$$[dx/dt]_2 = (f(t)/m) - (k/m)x_1 - (c/m)x_2 \quad,$$

$$x_1 = y \quad.$$

Hence we have the state-matrix equation choosing y as output:

$$\left\{ \begin{matrix} \dot{x_1} \\ \dot{x_2} \end{matrix} \right\} = \left\{ \begin{matrix} 0, & 1 \\ \frac{-k}{m}, & \frac{-c}{m} \end{matrix} \right\} \left\{ \begin{matrix} x_1 \\ x_2 \end{matrix} \right\} + \left\{ \begin{matrix} 0 \\ \frac{1}{m} \end{matrix} \right\} f(t)$$

$$y = [1,0] \left\{ \begin{matrix} x_1 \\ x_2 \end{matrix} \right\}$$

2. Simulation of the system assuming the following values for the problem variables and the coefficients (parameters):

magnitude scaling:

m = 10 kg;

c = 1000;

k = 600;

y = 1.0;

dy/dt = 4.0;

$|y_{max}|$ = 5cm;

$|[dy/dt]_{max}|$ = 50cm and f(t) = 10N.

Now since

$d^2y/dt^2 =$

(f(t)/m) - (k/m)y - (c/m)[dy/dt],

d([dy/dt])/dt = (f(t)/m) - (k/m)y

-(c/m)[dy/dt].

$(d([dy/dt]/[dy/dt]_{max})/dt)*$

$[dy/dt]_{max}| =$

$(f(t)/m)-(k/m)(|y/y_{max}|)*(y_{max})-$

$(c/m)|[dy/dt]/[dy/dt]_{max}|$

$*[dy/dt]_{max}$

Or,

$$\frac{d\left(\dot{y}\right)}{dt}$$

$$=$$

$$\frac{f(t)}{m} - \frac{ky}{m} - \frac{c\left(\dot{y}\right)}{m},$$

$$\frac{d\left(\dfrac{\dot{y}}{\dot{y}\max}\right)\dot{y}\max}{dt}$$

$$=$$

$$\frac{f(t)}{m} - \frac{k}{m}\left|\frac{y}{y\max}\right|y\max - \frac{c}{m}\left|\frac{\dot{y}}{\left(\dot{y}\right)\max}\right|\left(\dot{y}\right)\max$$

dividing all through by [dy/dt]_{max}:

```
dividing all through by [dy/dt]max:
d([dy/dt]/[dy/dt]max)/dt =
(1/[dy/dt]max)*(f(t)/m)-
(1/[dy/dt]max)*(k/m)(|y/ymax|)*(ymax)
-(1/[dy/dt]max)*
(c/m)|[dy/dt]/[dy/dt]max|*[dy/dt]max
```

Or,

234

$$\frac{d\left(\dfrac{\dot{y}}{\dot{y}\max}\right)}{dt} =$$

$$\frac{1}{\dot{y}\max}\frac{f(t)}{m} - \frac{k}{m}\left|\frac{y}{\dot{y}\max}\right|\frac{y\max}{\dot{y}\max} - \frac{c}{m}\left|\frac{\dot{y}}{\left(\dot{y}\right)\max}\right|\frac{\left(\dot{y}\right)\max}{\dot{y}\max}$$

Considering the coefficients in the magnitude scaled model we have:

1. $(K/m)*(1/[dy/dt]_{max})*(y_{max}) = (600)/10 * (5/50) = 6.0$,

2. $C/m = 1000/10 = 100$.

The second coefficient falls outside the limit $(0.1 \leq$ coefficient $\leq 10)$ thus we time scale. This will be done as below: multiplying the magnitude scale equation all thro (that is both sides) by 1/B we have:

1/B(d([dy/dt]/[dy/dt]ₘₐₓ)/dt)   =

(1/[dy/dt]ₘₐₓ)*(1/B)*(f(t)/m)-

(1/B)*(1/[dy/dt]ₘₐₓ)*(k/m)(|y/yₘₐₓ|)

(yₘₐₓ)-(1/B)(1/[dy/dt]ₘₐₓ)*

(c/m)|[dy/dt]/[dy/dt]ₘₐₓ|

*[dy/dt]ₘₐₓ

Or,

$$\frac{1}{\beta}\frac{d\left(\dfrac{\dot{y}}{\dot{y}\,max}\right)}{dt}=$$

$$\frac{1}{\dot{y}\,max}\frac{f(t)}{m\beta}-\frac{k}{m\beta}\left|\frac{y}{\dot{y}\,max}\right|\frac{y\,max}{\dot{y}\,max}-\frac{c}{m\beta}\left|\frac{\dot{y}}{\left(\dot{y}\right)max}\right|$$

Thus,

```
d([dy/dt]/[dy/dt]_max)/d(τ) =
(1/[dy/dt]_max)*(1/B)*(f(t)/m)          -
(1/B)*
(1/[dy/dt]_max)*(k/m)(|y/y_max|)*(y_max)
-(1/B)*
(1/[dy/dt]_max)*(c/m)|[dy/dt]/[dy/dt
]_max|*[dy/dt]_max.
```

Or,

$$\frac{d\left(\dfrac{\dot{y}}{\dot{y}\,\mathrm{max}}\right)}{d\tau} = \frac{1}{\dot{y}\,\mathrm{max}}\frac{f(t)}{m\beta} - \frac{k}{m\beta}\left|\frac{y}{y\,\mathrm{max}}\right|\frac{y\,\mathrm{max}}{\dot{y}\,\mathrm{max}} - \frac{c}{m\beta}\left|\frac{\dot{y}}{\left(\dot{y}\right)\mathrm{max}}\right|$$

We now have for the coefficients:

$(1/B) * (1/[dy/dt]_{max}) * (k/m) * (y_{max})$

$= ((600)/(10 * 10)) * (5/50)$

$= 0.6,$

2. $C/(mB) = 1000/(10 * 10) = 10$.

The program:

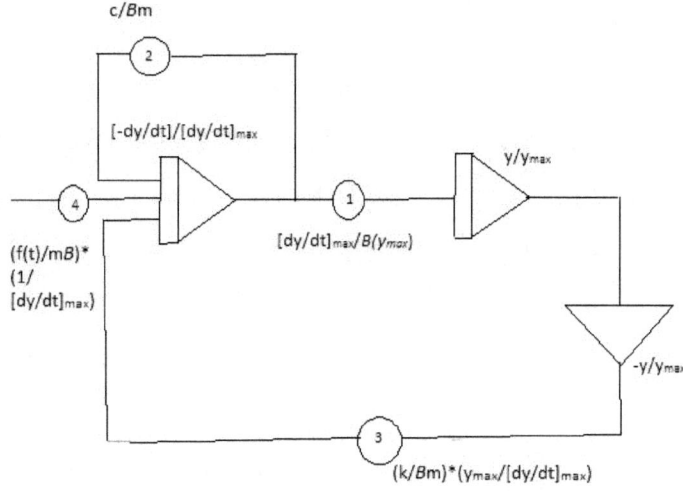

Static checking:

Problem variable	Maximum value	Computer variable (problem variable/(problem variable maximum value)) or amplifier output.
Y	5	y/5
[dy/dt]	50	[dy/dt]/50

Hence the program becomes:

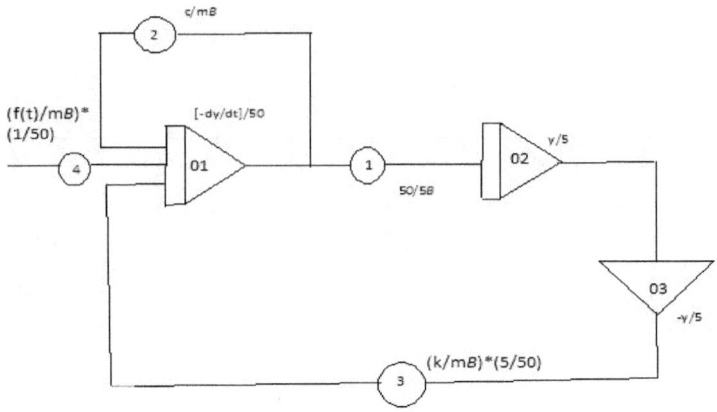

Calculation of outputs of amplifiers based on the program using the given values :

A01 = - [dy/dt]/50 = -4/50 ,

A02 = y/5 = 1/5 ,

A03(inverter) = -y/5 = -1/5

Calculation of integrator inputs based on the program :

input to integrator 02:

-[dy/dt]/50 * (50/5B) =

-[dy/dt]/5B = -4/(5 *10) = -4/50 ,

input to integrator 01: (A01 * Pot 2) + Pot 4 + (A03 * Pot 3) :

```
(-[dy/dt]/50*(c/mB)) +
((f(t)/mB))*(1/50)) +
(((-y)/5)*(5/50))*((k/mB)) ,
=  -4/50  *  (1000/(10  *  10))  +
(10/100)*(1/50)
+    (((-1/5)*(5/50)*(600/100)))
=   (-400 + 1 - 60)/500 = (-460 +
1)/500 = -459/500
```

Calculation of amplifier outputs
based on the equation:
first Amplifier output :
-[dy/dt]/50 = -4/50 ,
second Amplifier output :
y/5 = 1/5 ,
third Amplifier output (A03) :
-y/5 = -1/5
Calculation of integrator inputs
based on the equation : D01
(input to integrator 1) =
d([dy/dt]/50)/d(τ) =
1/B * (d([dy/dt]/50)/dt) =
1/B* (d([dy/dt]/50)/dt) =

$(1/50B)*(d([dy/dt])/dt)$ =

$(1/(50*B))*(d^2y/dt^2)$ =

$1/(50*B) * ((f(t)/m) - (k/m)*y - ((c/m)*([dy/dt])))$

= $1/(50*10) *((10/10) - ((600 * 1)/10) - ((1000 * 4)/ 10))$

= $(1/500) * (1 - 60 - 400)$

= $((1/500)*(-(- 1 + 60 + 400)))$

= $-459/500$,

$D02 = - [dy/dt]/(5B)$

= $-4 /(5 * 10) = -4/50$.

A comparism of the two different calculations shows that they are same.

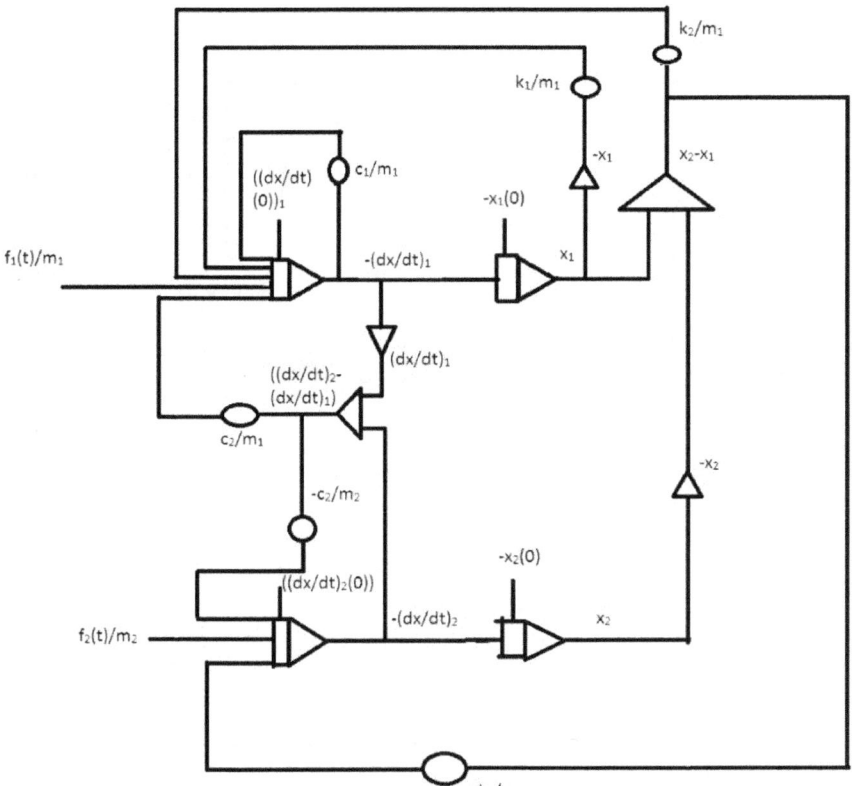

244

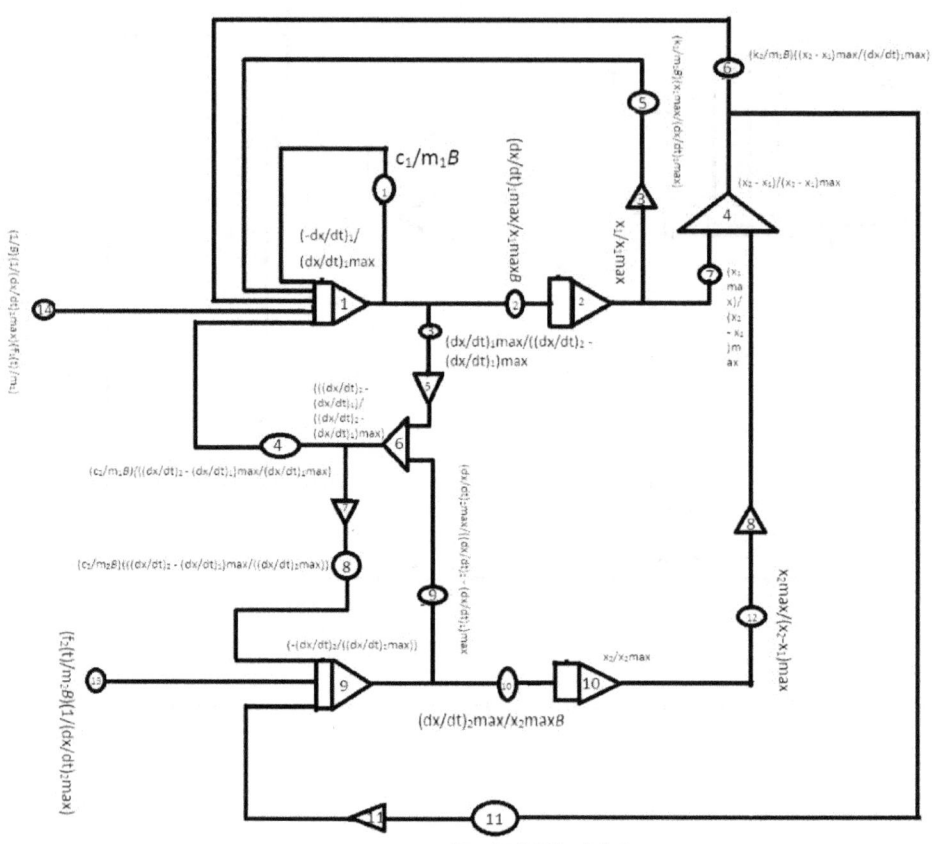

A := Amplifier, P := Pot.

System Component	Output Value
A1	$\left(\dfrac{-\dot{x}_1}{\left(\dot{x}_1\right)\max} \right)$
A2	$\left(\dfrac{x_1}{x_1 \max} \right)$
A3	-A2
A4	$\left(\dfrac{x_2 - x_1}{\left(x_2 - x_1\right)\max} \right)$
A5	-P3
A6	$\left(\dfrac{\left(\dot{x}_2\right) - \left(\dot{x}_1\right)}{\left(\left(\dot{x}_2\right) - \left(\dot{x}_1\right)\right)\max} \right)$
A7	-A6

A8	-P12
A9	$\left(\dfrac{-\left(\dot{x}_2 \right)}{\left(\dot{x}_2 \right) \max} \right)$
A10	$\dfrac{x_2}{x_2 \max}$
A11	-P11
P1	$\dfrac{c_1}{m_1 \beta}$
P2	$\dfrac{\left(\dot{x}_1 \right) \max}{x_1 \max \beta}$
P3	$\dfrac{\left(\dot{x}_1 \right) \max}{\left(\left(\dot{x}_2 \right) - \left(\dot{x}_1 \right) \right) \max}$
P4	$\left(\dfrac{c_2}{m_1 \beta} \right) \dfrac{\left(\left(\dot{x}_2 \right) - \left(\dot{x}_1 \right) \right) \max}{\left(\dot{x}_1 \right) \max}$

P5	$$\left(\dfrac{k_1}{m_1\beta}\right)\dfrac{x_1\max}{\left(\overset{\bullet}{x_1}\right)\max}$$
P6	$$\left(\dfrac{k_2}{m_1\beta}\right)\dfrac{(x_2-x_1)\max}{\left(\overset{\bullet}{x_1}\right)\max}$$
P7	$$\dfrac{x_1\max}{(x_2-x_1)\max}$$
P8	$$\left(\dfrac{c_2}{m_2\beta}\right)\left(\dfrac{(\left(\overset{\bullet}{x_2}\right)-\left(\overset{\bullet}{x_1}\right))\max}{\left(\overset{\bullet}{x_2}\right)\max}\right)$$
P9	$$\left(\dfrac{\left(\overset{\bullet}{x_2}\right)\max}{(\left(\overset{\bullet}{x_2}\right)-\left(\overset{\bullet}{x_1}\right))\max}\right)$$
P10	$$\left(\dfrac{\left(\overset{\bullet}{x_2}\right)\max}{(x_2)\max\,\beta}\right)$$
P11	$$\left(\dfrac{(x_2-x_1)\max}{\left(\overset{\bullet}{x_2}\right)\max}\right)\dfrac{k_2}{m_2\beta}$$

P12	$\left(\dfrac{(x_2)\max}{(x_2 - x_1)\max} \right)$
P13	$\dfrac{f_2(t)}{m_2 \beta} \left(\dfrac{1}{\left(\dot{x}_2 \right)\max} \right)$
P14	$\dfrac{f_1(t)}{m_1 \beta} \left(\dfrac{1}{\left(\dot{x}_1 \right)\max} \right)$

REFERENCE

1. Katsuhiko Ogata. Modern Control Engineering. 4th edition. India - university of minnesota : Pearson (Prentice Hall), 2008.

2. Stanley Shinners M. Modern Control System theory and application. 2nd edition. Canada :addison-wesley Publishing company, 1980.

3. Gene F. and David J. P. Digital Control of Dynamic Systems. Canada: addison-wesley Publishing company, 1980.

4. Mathew Boelkins R., et al. Differential Equations with linear Algebra. New York: oxford university Press, 2009.

5. Hung and Ramin. Dynamic Systems: modelling and Analysis. USA: Irwin/McGraw hill, 1997.

6. Fred Ricci J. Analog/logic computer Programming and simulation. USA. Spartan books ,

1972.

7. George Simmons F. Differential equations with Applications and historical notes. 2nd edition. Singapore:McGraw-Hill, 1991.

8. Anthony croft, et al. Engineering Mathematics, a foundation for electronic, electrical, communications and Systems Engineers. 4th edition. UK: Pearson, 2013.

9. Clares and George. Continuous and Discrete Signal and System Analysis. USA:Holt,Rinehart and Winston Inc ,1974.

10. Jacob M. and Christos C. H. Integrated Electronics: analog and digital circuits and systems. Singapore: McGraw-Hill book Company, 1992.

Index

Homogenous equations , 44 , 47

I

Independent variable , 4 , 5 , 45

Integrating factor, 33, 35, 66

Integration by part method, 41

Indices, 59

Inverse Laplace transform, 88 , 115

Infinity, 104

Initial and final value theorems , 124, 125

Initial value theorem, 125

Input, 141 , 142 , 143

initial condition , 171

K

Kirchhoff voltage laws , 172

L

Linear differential equations, 32

Laplace transform, 88, 89

Laplace transform of derivatives, 105

Laplace transform of integral, 110

law of conservation of mass , 172

M

Method of variation of parameters,

For I was my father's son...He taught me also, and said unto me, let thine heart retain my words....Get wisdom, get understanding: forget it not.....
Proverb 4 vs 3 – 5(AKJV)